PRINCIPLE OF RELATIVITY

PRINCIPLE OF RELATIVITY

A. N. Whitehead, Sc.D., F. R. S.
Hon. D.Sc. (Manchester), Hon. LL.D. (St Andrews) Fellow of Trinity College, Cambridge, and Professor of Applied Mathematics in the Imperial College of Science and Technology

Introduction by Amit Hagar

Barnes & Noble Books
New York

Originally published in 1922

This edition published by Barnes & Noble Publishing, Inc.

Cover Design by Stacey May

2005 Barnes & Noble Publishing, Inc.

ISBN 0-7607-6521-9

Printed and bound in the United States of America

1 3 5 7 9 10 8 6 4 2

To my Wife
Whose encouragement and counsel have
made my life's work possible

CONTENTS

INTRODUCTION

INTERWEAVING science and metaphysics, *The Principle of Relativity* presents what is commonly acknowledged as the most interesting alternative to Einstein's general theory of relativity (GTR). The great mathematician and philosopher Alfred North Whitehead furthermore spells out his view of geometry, spacetime, and Nature in this intriguing book. Originally published in 1922, the book offers a different paradigm from Einstein's, elegant and simple in its mathematical formulation with its own philosophical background and agenda. Few other books exemplify as it does the intricate inter-relations between physics and philosophy. Scholars from both domains shall find here much to chew on, as well as the layman who is interested in the history of ideas of the twentieth century.

The youngest of four children of an Anglican vicar, Whitehead (1861-1947)—one of the most interesting and imaginative scholars of our era—showed no sign of the genius that he was later in life. As a scholar, his academic interests spanned mathematics, science, and metaphysics, all of which were thoroughly and carefully treated by him with a unique style. Guided by the intellectual honesty and the personal touch that is so characteristic of his writings, only few follow Whitehead in the rarely taken path that combines physics and philosophy. His ingenuity and creativity shall remain an inspiration to generations of scholars.

After winning two scholarships for studying mathematics in Trinity College, Cambridge, Whitehead was offered a fellowship as an assistant lecturer there. Despite a poor publication record he was soon promoted and became a lecturer. The shift of emphasis in his career, from teaching to publishing, came with his marriage in late 1890. It was also marked by his decision to renounce Christianity. He himself stated that the biggest factor in his becoming an agnostic was the rapid developments in science; particularly his view that Newton's physics was false. It may seem surprising to many that the correctness of Newton's physics could be a major factor in deciding anyone's religious views. However, one has to understand the complex person that Whitehead was and, in particular, the interest which he was developing in philosophy and metaphysics.

Whitehead left Cambridge in 1910 and went to London, and then to Harvard, where he was the chair of the philosophy department until his retirement. Apart from his metaphysics, he is perhaps best known for his collaboration with Bertrand Russell, who came to Cambridge in 1890 as an undergraduate and was immediately spotted by the talented lecturer. A decade later, the student and the master began collaborating on one of the most ambitious projects in the philosophy of mathematics, *Principia Mathematica,* which was an attempt to supply mathematics with rigorous logical foundations. When the first volume of this monumental work was finished, Whitehead and Russell began to go their separate ways. Perhaps inevitably, Russell's anti-war activities during World War I, in which Whitehead lost his youngest son, also led to something of a split between the two men. Nevertheless, they remained on relatively good terms for the rest of their lives. It was then that Whitehead turned his attention to the philosophy of science. This interest arose out of the attempt to explain the relation of formal mathematical theories in physics to their basis in experience and was sparked by the revolution brought on by Einstein's GTR, to which he had developed an alternative.

Whitehead's theory of gravity has been called "a thorn in Einstein's side." Since it agrees with GTR in its prediction for all the classical tests, the real issues between Whitehead and Einstein are not physical but philosophical. The theory is closely connected with Whitehead's philosophy of nature and with his metaphysics, which are spelled out in his *The Concept of Nature* (1920), *Science and the Modern World* (1925), and *Process and Reality* (1928). Its popular debut was in *The London Times Educational Supplement* in January 1920, shortly after Arthur Eddington, the famous British astrophysicist, verified GTR's predictions about "the bending of light" in the vicinity of the sun during the 1919 eclipse. Eddington himself proved a few years later that Whitehead's alternative has the same solution as Einstein's in the special case of the stationary gravitational field due to a single mass point. This meant that it was indeed empirically equivalent to GTR, at least with respect to the standard tests that GTR has passed so far, e.g., the perihelion precession of Mercury and the bending of light rays.

Later on in the 1920s, the comparison between the two rival theories was conducted mainly on the level of conceptual analysis. With the shift of interest among physicists to the realms of quantum mechanics and nuclear physics during the 1930s and the 1940s, Whitehead, by then regarded as a metaphysician, was ignored. The re-evaluation of Whitehead's ideas began in the 1950s due to an Irish physicist, Synge, who esteemed Whitehead's theory for its elegance and originality. Setting aside Whitehead's philosophical agenda, Synge reconstructed Whitehead's mathematical formulae in Einstein's terminology to make them accessible to contemporary physicists. In the 1960s, there was a revival of empirical tests for gravitational theories, especially due to the rapid progress in technology and astronomy. In 1965, a series of experiments were conducted measuring the galactic red shift (an effect similar to the Doppler effect due to the expansion of the universe). The accuracy of this experiment was about twenty times higher than previous astronomical observations and was proven to be a strong support to Einstein's GTR. Whitehead's theory in its original form predicted a gravitational red shift slightly different than that of Einstein, but qualitatively the two rival theories were still empirically equivalent.

It was only in the early 1970s when the American astrophysicist Clifford M. Will claimed to have shown that Whitehead's theory should be in conflict with experimental data with respect to certain geophysical effects. Yet also here it was pointed out that the alleged refutation of Whitehead's theory on geophysical grounds relied on a certain astronomical assumption that was probably false. This loophole did not necessarily resurrect Whitehead's theory, but it did imply that there is interesting work yet to be done.

Einstein presented his special theory of relativity as stemming from two phenomenological principles: the principle of relativity and the principle of the constancy of the velocity of light in vacuum. Based on these two principles one can, as Minkowski did, construct spacetime as a four-dimensional manifold equipped with a Minkowskian (flat) metric and extract the Lorentz transformations between different inertial reference frames. To these principles Einstein added the equivalence principle between gravitation and inertia and generalized his special theory to the case of non-inertial frames. The consequences of Einstein's GTR were far-reaching: Physical objects such as electromagnetic and matter fields were now treated as geometrical objects, and the geometry of spacetime was fixed by the distribution of matter in it. With the absence of matter one should expect spacetime to be flat, or Euclidean; with its presence — curved, or non-Euclidean.

Whitehead's theory differs from Einstein's in its physical content as well as in its philosophical background. Whitehead rejected Einstein's equivalence principle and proposed a different reconstruction of the Minkowskian four-dimensional manifold based on weaker principles than Einstein's. His main objection was to the idea that spacetime can change its geometry contingent upon the presence of matter. To Whitehead, who believed in the uniformity of Nature, the idea of variable curvature implied the precedence of matter over spacetime, and as such had no room in the metaphysics he was developing. As he stresses in the preface to *The Principle of Relativity*, it is the *uniformity* which is crucial to his outlook, and not the specific geometry of spacetime. The former is a necessary condition — a guiding principle for doing physics, indeed for conducting science in general — while the latter is open to empirical revisions: Spacetime could possess either flat or curved geometry, according to Whitehead, as long as it is uniform.

Whitehead's metaphysics, also known as Process Philosophy, is as original as his physics. In its core are the ideas of the priority of events over objects and the precedence of becoming over being. Material objects, in this view, are not subjects of predicates or bearers of properties, but adjectives of events; they do not affect spacetime structure, rather they are affected by it and merely represent it. Relying on this priority, and combining it with an epistemological argument, Whitehead deduced the idea that Nature must be uniform, and that all events are interconnected through a relation of relatedness.

This metaphysics resulted in an alternative theory of gravitation that describes gravity as a retarded action-at-a-distance force and in a spacetime metric that describes the uniform dynamic relatedness of events in process. Whitehead's metric, as Jonathan Bain, a contemporary American philosopher, nicely puts it in a recent article, ". . .is absolute insofar as this dynamic relatedness is uniform and exists in causal independence of matter fields in order for knowledge and induction to be possible. . . ," and ". . .is dynamic insofar as it describes not a fixed eternally existing set of substantival spacetime points, but rather an ever-changing flux of interrelated events." The title of the book is thus a misnomer: It should have been *The Principle of Relatedness.*

Thought-provoking and genuine, Whitehead's *The Principle of Relativity* stands as one of the classics in the philosophy of spacetime physics, and the ideas it propounds continue to attract and inflame physicists and philosophers even today. Referring as he was to Einstein, Whitehead himself said in the book that "the worst homage we can pay to genius is to accept uncritically formulations of truths which we owe to it." This kind of spirit applies in the case of Whitehead as well.

Amit Hagar is a philosopher of physics with a Ph.D. from the University of British Columbia, Vancouver. His area of specialization is the conceptual foundations of modern physics, especially in the domains of statistical and quantum mechanics.

PREFACE

THE present work is an exposition of an alternative rendering of the theory of relativity. It takes its rise from that 'awakening from dogmatic slumber'—to use Kant's phrase—which we owe to Einstein and Minkowski. But it is not an attempt to expound either Einstein's earlier or his later theory. The metrical formulae finally arrived at are those of the earlier theory, but the meanings ascribed to the algebraic symbols are entirely different. As the result of a consideration of the character of our knowledge in general, and of our knowledge of nature in particular, undertaken in Part I of this book and in my two previous works[1] on this subject, I deduce that our experience requires and exhibits a basis of uniformity, and that in the case of nature this basis exhibits itself as the uniformity of spatio-temporal relations. This conclusion entirely cuts away the casual heterogeneity of these relations which is the essential of Einstein's later theory. It is this uniformity which is essential to my outlook, and not the Euclidean geometry which I adopt as lending itself to the simplest exposition of the facts of nature. I should be very willing to believe that each permanent space is either uniformly elliptic or uniformly hyperbolic, if any observations are more simply explained by such a hypothesis.

It is inherent in my theory to maintain the old division between physics and geometry. Physics is the science of the contingent relations of nature and geometry expresses its uniform relatedness.

The book is divided into three parts. Part I is concerned with general principles and may roughly be described as mainly philosophical in character. Part II is devoted to the physical applications and deals with the particular results deducible from the formulae assumed for the gravitational and electromagnetic fields. In relation to the spectral lines these formulae would require a 'limb effect' and a duplication or a triplication of individual lines, analogous to phenomena already observed. Part III is an exposition of the elementary theory of tensors. This Part has been added for one reason because it may be useful to many mathematicians who may be puzzled by some of the formulae and procedures of Part II. But this Part is also required by another reason. The theory of tensors is usually expounded under the guise of geometrical metaphors which entirely mask the type of application which I give to it in this work. For example, the whole idea of any 'fundamental tensor' is foreign to my purpose and impedes the comprehension of my applications.

The order in which the parts should be studied will depend upon the psychology of the reader. I have placed them in the order natural to my own mind, namely, general principles, particular applications, and finally the general exposition of the mathematical theory of which special examples have occurred in the discussion of the applications. But a physicist may prefer to start with Part II, referring back to a few formulae which have been mentioned at the end of Part I, and a mathematician may start with Part III. The whole evidence requires a consideration of the three Parts.

Practically the whole of the book has been delivered in the form, of lectures either in America at the College of Bryn Mawr, or before the Royal Society of Edinburgh, or to my pupils in the Imperial College. I have carefully preserved the lecture form and also some reduplication of statement, particularly in Part I.

The exposition of a novel idea which has many reactions upon diverse current modes of thought is a difficult business. The most successful example in the history of science is, I think,

Galileo's 'Dialogues on the Two Systems of the World.' An examination of that masterly work will show that the dialogue form is an essential element to its excellence. It allows the main expositor of the dialogues continually to restate his ideas in reference to diverse trains of thought suggested by the other interlocutors. Now the process of understanding new conceptions is essentially the process of laying the new ideas alongside of our pre-existing trains of thought. Accordingly for an author of adequate literary ability the dialogue is the natural literary form for the prolonged explanation of a tangled subject. The custom of modern presentations of science, and my own diffidence of success in the art of managing a dialogue, have led me to adopt the modified form of lectures in which the audiences—real audiences, either in America, Edinburgh or South Kensington—are to be regarded as silent interlocutors demanding explanations of the various aspects of the theory.

Chapter II was originally delivered[2] in Edinburgh as a lecture to the Royal Society of Edinburgh when it did me the honour of making me the first recipient of the 'James-Scott Prize' for the encouragement of the philosophy of science. Chapter IV was originally delivered[3] at the College of Bryn Mawr, near Philadelphia, on the occasion of a festival promoted by the former pupils and colleagues of Prof. Charlotte Angus Scott in honour of her work as Professor of Mathematics at the college since its foundation.

My thanks are due to my colleague, Assistant-Professor Sillick, for the figure on p. 31. I am also further indebted to him for a series of beautiful slides containing the mathematical formulae of Chapter IV; even the admirable printing of the Cambridge University Press will not compensate readers of this book for the loss of the slides as used in the original lecture.

In acknowledging my obligations to the efficiency and courtesy of the staff of the University Press, I take the opportunity of paying a respectful tribute to the work of the late Mr A. R. Waller as secretary of the Press Syndicate. The initial negotiations respecting this book were conducted through him and he died just as the printing commenced. The loss of his wisdom, his knowledge, and his charm will leave a gap in the hearts of all those who have to deal with the great Institution which he served so well.

A. N. W.

15 *September*, 1922.

PART I

GENERAL PRINCIPLES

CHAPTER I

PREFATORY EXPLANATIONS

THE doctrine of relativity affects every branch of natural science, not excluding the biological sciences. In general, however, this impact of the new doctrine on the older sciences lies in the future and will disclose itself in ways not yet apparent. Relativity, in the form of novel formulae relating time and space, first developed in connection with electromagnetism, including light phenomena. Einstein then proceeded to show its bearing on the formulae for gravitation. It so happens therefore that owing to the circumstances of its origin a very general doctrine is linked with two special applications.

In this procedure science is evolving according to its usual mode. In that atmosphere of thought doctrines are valued for their utility as instruments of research. Only one question is asked: Has the doctrine a precise application to a variety of particular circumstances so as to determine the exact phenomena which should be then observed? In the comparative absence of these applications beauty, generality, or even truth, will not save a doctrine from neglect in scientific thought. With them, it will be absorbed.

Accordingly a new scientific outlook clings to those fields where its first applications are to be found. They are its title deeds for consideration. But in testing its truth, if the theory have the width and depth which marks a fundamental reorganisation, we cannot wisely confine ourselves solely to the consideration of a few happy applications. The history of science is strewn with the happy applications of discarded theories. There are two gauges through which every theory must pass. There is the broad gauge which tests its consonance with the general character of our direct experience, and there is the narrow gauge which is that mentioned above as being the habitual working gauge of science. These reflections have been suggested by the advice received from two distinguished persons to whom at different times I had explained the scheme of this book. The philosopher advised me to omit the mathematics, and the mathematician urged the cutting out of the philosophy. At the moment I was persuaded: it certainly is a nuisance for philosophers to be worried with applied mathematics, and for mathematicians to be saddled with philosophy. But further reflection has made me retain my original plan. The difficulty is inherent in the subject matter.

To expect to reorganise our ideas of Time, Space, and Measurement without some discussion which must be ranked as philosophical is to neglect the teaching of history and the inherent probabilities of the subject. On the other hand no reorganisation of these ideas can command confidence unless it supplies science with added power in the analysis of phenomena. The evidence is two-fold, and is fatally weakened if the two parts are disjoined.

At the same time it is well to understand the limitations to the meaning of 'philosophy' in this connection. It has nothing to do with ethics or theology or the theory of aesthetics. It is solely engaged in determining the most general conceptions which apply to things observed by the senses. Accordingly it is not even metaphysics: it should be called pan-physics. Its task is to formulate those principles of science which are employed equally in every branch of natural science. Sir J. J. Thomson, reviewing in *Nature*[1] Poynting's *Collected Papers,* has quoted a statement taken from one of Poynting's addresses:

'I have no doubt whatever that our ultimate aim must be to describe the sensible in terms of the sensible.'

Adherence to this aphorism, sanctioned by the authority of two great English physicists, is the keynote of everything in the following chapters. The philosophy of science is the endeavour to formulate the most general characters of things observed. These sought-for characters are to be no fancy characters of a fairy tale enacted behind the scenes. They must be observed characters of things observed. Nature is what is observed, and the ether is an observed character of things observed. Thus the philosophy of science only differs from any of the special natural sciences by the fact that it is natural science at the stage before it is convenient to split it up into its various branches. This philosophy exists because there is something to be said before we commence the process of differentiation. It is true that in human thought the particular precedes the general. Accordingly the philosophy will not advance until the branches of science have made independent progress. Philosophy then appears as a criticism and a corrective, and—what is now to the purpose—as an additional source of evidence in times of fundamental reorganisation.

This assignment of the rôle of philosophy is borne out by history. It is not true that science has advanced in disregard of any general discussion of the character of the universe. The scientists of the Renaissance and their immediate successors of the seventeenth century, to whom we owe our traditional concepts, inherited from Plato, Aristotle and the medieval scholastics. It is true that the New Learning reacted violently against the schoolmen who were their immediate predecessors; but, like the Israelites when they fled from Egypt, they borrowed their valuables—and in this case the valuables were certain root-presuppositions respecting space, time, matter, predicate and subject, and logic in general. It is legitimate (as a practical counsel in the management of a short life) to abstain from the criticism of scientific foundations so long as the superstructure 'works.' But to neglect philosophy when engaged in the re-formation of ideas is to assume the absolute correctness of the chance philosophic prejudices imbibed from a nurse or a schoolmaster or current modes of expression. It is to enact the part of those who thank Providence that they have been saved from the perplexities of religious enquiry by the happiness of birth in the true faith. The truth is that your available concepts depend upon your philosophy. An examination of the writings of John Stuart Mill and his immediate successors on the procedure of science—writings of the highest excellence within their limitations—will show that they are exclusively considering the procedure of science in the framing of laws with the employment of given concepts. If this limitation be admitted, the conclusion at once follows that philosophy is useless in the progress of science. But when once you tamper with your basic concepts, philosophy is merely the marshalling of one main source of evidence, and cannot be neglected.

But when all has been said respecting the importance of philosophy for the discovery of scientific truth, the narrow-gauged pragmatic test will remain the final arbiter. Accordingly I now proceed to a summary account of the general doctrine either implicit or explicit in the following pages or in my two previous books[2] on this subject, and to detail the facts of experience which receive their explanation from it or should be observed if it be true.

A relativistic view of time is adopted so that an instantaneous moment of time is nothing else than an instantaneous and simultaneous spread of the events of the universe. But in the concept of instantaneousness the concept of the passage of time has been lost. Events essentially involve this passage. Accordingly the self-contradictory idea of an instantaneous event has to be replaced by that of an instantaneous configuration of the universe. But what is directly observed is an event. Thus a duration, which is a slab of time with temporal thickness, is the final fact of observation from which moments and configurations are deduced as a limit which is a logical ideal of the exact precision inherent in nature. This process of deducing limits is considered in detail in my two previous books under the title Extensive Abstraction. But it is an essential assumption that a concrete fact of nature always includes temporal passage.

A moment expresses the spread of nature as a configuration in an instantaneous three dimensional space. The flow of time means the succession of moments, and this succession includes the whole of nature. Rest and motion are direct facts of observation concerning the relation of objects to the durations whose limits are the moments of this flow of time. By means of rest a permanent point is defined which is merely a track of event-particles with one event-particle in every moment.

Refined observation (in the form of the Michelson-Morley experiment and allied experiments) shows that there are alternative flows of time—or time-systems, as they will be called,—and that the time-system actually observed is that one for which (roughly speaking) our body is at rest. Accordingly in different circumstances of motion, space and time mean different things, the moments of one time-system are different from the moments of another time-system, the permanent points of one time-system are different from those of another time-system, so that the permanent space of one time-system is distinct from the permanent space of another time-system.

The properties of time and space express the basis of uniformity in nature which is essential for our knowledge of nature as a coherent system. The physical field expresses the unessential uniformities regulating the contingency of appearance. In a fuller consideration of experience they may exhibit themselves as essential; but if we limit ourselves to nature there is no essential reason for the particular nexus of appearance.

Thus times and spaces are uniform.

Position in space is merely the expression of diversity of relations to alternative time-systems. Order in space is merely the reflection into the space of one time-system of the time-orders of alternative time-systems.

A plane in space expresses the quality of the locus of intersection of a moment of the time-system in question (call it 'time-system *A*') with a moment of another time-system (time-system *B*).

The parallelism of planes in the space of time-system *A* means that these planes result from the intersections of moments of *A* with moments of one other time-system *B*.

A straight line in the space of time-system *A* perpendicular to the planes due to time-system *B* is the track in the space of time-system *A* of a body at rest in the space of time-system *B*.

Thus the uniform Euclidean geometry of spaces, planeness, parallelism, and perpendicularity are merely expressive of the relations to each other of alternative time-systems.

The tracks which are the permanent points of the same time-system are also reckoned as parallels.

Congruence—and thence, spatial measurement—is defined in terms of the properties of parallelograms and the symmetry of perpendicularity.

Accordingly, position, planes, straight lines, parallelism, perpendicularity, and congruence are expressive of the mutual relations of alternative time-systems.

The symmetrical properties of relative velocity are shown (in *The Principles of Natural Knowledge*) to issue in a critical velocity *c*, which thus is defined without reference to the velocity of light. However experiment shows that for our purposes it must be a near approach to that velocity. The final result is the geometry and kinematic which are explained in Chapter IV of the present volume.

A physical object, such as a mass-particle or an electron, expresses the character of the future so far as it is determined by the happenings of the present. The exact meaning of an object as an entity implicated in events is explained. The track of an object amid events is determined by the 'stationary' property of the impetus realised by the pervasion of the track by the object. This impetus depends partly on the intrinsic character of the object—

e.g. its mass or its electric charge—and partly on the intrinsic potential impetus of the track itself. This potential impetus arises from the physical character of the events of the region due to the presence of other objects in the past. This physical character is partly gravitational and partly electrical.

This dependence of physical character on antecedent objects is directly expressed by the formula here adopted for the gravitational law. This law also gives the most direct expression to the principle that the flux of time is essential to the concrete reality of nature, so that a loss of time-flux means a transference to a higher abstraction. It gives this expression by conceiving the attracting body as pervading an element of its track and not as at an event-particle. This law gives the Einstein expression for the revolution of the perihelion of mercury.

The electromagnetic equations adopted are Maxwell's equations modified by the gravitational tensor components in the well-known way. Light is given no privileged position, and all deductions concerning light follow directly from treating it as consisting of short waves of electromagnetic disturbance. In this way Einstein's assumption that a ray of light follows the path

$$dJ^2 = 0$$

[i.e. in Einstein's notation

$$ds^2 = 0]$$

can be proved as an approximation due to the shortness of the waves.

The bending of the light rays in a gravitational field then follows.

With regard to the shift of spectral lines, there are three effects to be considered: (i) Einstein's predicted shift due to the gravitational potential, (ii) the limb effect which has been observed in the case of light from the sun, (iii) the doubling or trebling of spectral lines observed in the spectra due to some nebulae. Neither of the effects (ii) or (iii) has hitherto been explained.

As to (i) this is traced to the combination of two causes, one being the change in the apparent mass due to the gravitational potential and the other being the change in the electric cohesive forces of the molecule due to the gravitational field. The total result is that the period of vibration is changed from T to $T+\delta T$, where

$$\frac{\delta T}{T}=\frac{7}{6}\frac{\psi_4}{c^2},$$

ψ_4 being the gravitational potential. Einstein's result is ψ_4/c^2, so that the two formulae are practically identical for observational purposes.

With regard to effects (ii) and (iii) reasons are given, for believing that the molecules will separate into three groups sending a distant observer light of changing relative intensities as we pass from the centre of the disc of the emitting body (sun or nebulae) to the edge. One group has the above-mentioned shift, another has the shift

$$\frac{\delta T}{T}=\frac{3-2\eta}{2c^2}\psi_4$$

(where η is probably about 1/10, but may be nearly 1/5), and the third group has the shift

$$\frac{\delta T}{T}=\frac{2+\eta}{2c^2}\psi_4.$$

Under circumstances such that all or two of the groups send separately observable light, the trebling or doubling effects are explained to the extent of demonstrating the existence of causes for the multiplication of lines, other than those due to the motions of the matter of the nebulae. Under other circumstances (e. g. light from the sun's disc) in which the influence of the grouping is effective but not separately observable the shift approximates to

$$\frac{\delta T}{T}=\frac{1}{c^2}\psi_4\left\{1+\tfrac{1}{2}\eta+\tfrac{1}{4}(1-3\eta)\sin^2\beta_1\right\},$$

where β_1 varies from zero at the centre of the disc of the sun to $\pi/2$ at its edge. But there will be various intermediate circumstances between these extreme assumptions as to the observability of the grouping effect.

Finally in a steady electromagnetic field the electromagnetic equations predict two novel magnetic forces due to the gravitational field. These forces are excessively small: (i) A steady electric force at a point on the earth's surface (F in electrostatic units) should be accompanied by the horizontal magnetic force

$$1{\cdot}2 \times 10^{-9} \times F \sin \alpha \ (\text{gausses})$$

perpendicular to its direction and to the vertical, where α is the angle between these directions.

(ii) A steady current (I in electromagnetic measure) in a straight wire making an angle β with the vertical should produce at a point distant R from the wire the parallel magnetic force (i.e. in a direction parallel to the wire),

$$\frac{1}{2} \times 10^{-9} \times \cos \phi \sin 2\beta \times \frac{2I}{R} (\text{gausses}),$$

where Ø is the angle between the vertical plane through the wire and the plane through the wire and the point. The temperature of an attracting body should augment its gravitational field by an amount which is probably outside the limits of our observational powers.

CHAPTER II

THE RELATEDNESS OF NATURE

"Threads and floating wisps
Of being, ..."

CLEMENCE DANE'S *Will Shakespeare,* Act I.

YOU have conferred upon me the honour of becoming the first recipient of the 'James-Scott Prize,' and have at the same time assigned to me the duty of delivering a lecture upon the subject which this prize is designed to foster. In choosing the topic of a lecture which is to be the first of a series upon the philosophy of science, it seems suitable to explore the broadest possible aspect of the subject. Accordingly I propose to address you upon Relatedness and, in particular, upon the Relatedness of Nature. I feel some natural diffidence in speaking upon this theme in the capital of British metaphysics, haunted by the shade of Hume. This great thinker made short work of the theory of the relatedness of nature as it existed in the current philosophy of his time. It is hardly too much to say that the course of subsequent philosophy, including even Hume's own later writings and the British Empirical School, but still more in the stream which descends through Kant, Hegel and Caird, has been an endeavour to restore some theory of relatedness to replace the one demolished by Hume's youthful scepticism. If you once conceive fundamental fact as a multiplicity of subjects qualified by predicates, you must fail to give a coherent account of experience. The disjunction of subjects is the presupposition from which you start, and you can only account for conjunctive relations by some fallacious sleight of hand, such as Leibniz's metaphor of his monads engaged in mirroring. The alternative philosophic position must commence with denouncing the whole idea of 'subject qualified by predicate' as a trap set for philosophers by the syntax of language. The conclusion which I shall wish to enforce is that we can discern in nature a ground of uniformity, of which the more far-reaching example is the uniformity of space-time and the more limited example is what is usually known under the title, The Uniformity of Nature. My arguments must be based upon considerations of the utmost generality untouched by the peculiar features of any particular natural science. It is therefore inevitable that at the beginning my exposition will suffer from the vagueness which clings to generality.

Fact is a relationship of factors. Every factor of fact essentially refers to its relationships within fact. Apart from this reference it is not itself. Thus every factor of fact has fact for its background, and refers to fact in a way peculiar to itself.

I shall use the term 'awareness' for consciousness of factors within fact. A converse mode of statement is that awareness is consciousness of fact as involving factors. Awareness is itself a factor within fact.

I shall use the term 'cogitation' for consciousness of factors prescinded from their background of fact. It is the consciousness of the individuality of factors, in that each factor is itself and not another. A factor cogitated upon as individual will be called an 'entity.' The essence of cogitation is consciousness of diversity. The prescinding from the background of fact consists in limiting consciousness to awareness of the contrast of factors. Cogitation thus presupposes awareness and is limited by the limitations of awareness. It is the refinement of awareness, and the unity of consciousness lies in this dependence of cogitation upon awareness. Thus awareness is crude consciousness and cogitation is refined consciousness. For awareness all relations between factors are internal and for cogitation all relations between entities are external.

Fact in its totality is not an entity for cogitation, since it has no individuality by its reference to anything other than itself. It is not a relatum in the relationship of contrast. I might have used the term 'totality' instead of 'fact'; but 'fact' is shorter and gives rise to the convenient term 'factor.' Fact enters consciousness in a way peculiar to itself. It is not the sum of factors; it is rather the concreteness (or, embeddedness) of factors, and the concreteness of an inexhaustible relatedness among inexhaustible relata. If for one moment I may use the inadmissible word 'Factuality,' it is in some ways better either than 'fact' or 'totality' for the expression of my meaning. For 'fact' suggests one fact among others. This is not what I mean, and is a subordinate meaning which I express by 'factor.' Also 'totality' suggests a definite aggregate which is all that there is, and which can be constructed as the sum of all subordinate aggregates. I deny this view of factuality. For example, in the very conception of the addition of subordinate aggregates, the concept of the addition is omitted although this concept is itself a factor of factuality. Thus inexhaustibleness is the prime character of factuality as disclosed in awareness; that is to say, factuality (even as in individual awareness) cannot be exhausted by any definite class of factors. After this explanation I will now relapse into the use of 'fact' in the sense of 'factuality.'

The finiteness of consciousness, the factorisation of fact, the individualisation of entities in cogitation, and the opposition of abstract to concrete are all exhibitions of the same truth of the existence of limitation within fact. The abstract is a limitation within the concrete, the entity is a limitation within totality, the factor is a limitation within fact, and consciousness by its reference to its own standpoint within fact limits fact to fact as apprehended in consciousness. The treatment of the whole theory of limitation has suffered by the introduction of metaphors derived from a highly particular form of it, namely, derived from the analogy between extended things, such as that of whole to part and that of things mutually external to each other.

I use the term 'limitation' for the most general conception of finitude. In a somewhat more restricted sense Bergson uses the very convenient term 'canalisation.' This Bergsonian term is a useful one to keep in mind as a corrective to the misleading associations of the terms 'external' and 'internal,' or of the terms 'whole' and 'part.' It adds also a content to the negative term 'limitation.' Thus a factor is a limitation of fact in the sense that a factor refers to fact canalised into a system of relata to itself, i.e. to the factor in question. The mere negative limitation, or finitude, involved in a factor is exhibited in cogitation, wherein the factor degenerates into an entity and the canalisation degenerates into a bundle of external relations.

Thus also finite consciousness is a limitation of fact, in the sense that it is a factor canalising fact in ways peculiar to itself. We must get rid of the notion of consciousness as a little box with some things inside it. A better metaphor is that of the contact of consciousness with other factors, which is practically Hume's metaphor 'impression.' But this metaphor erroneously presupposes that fact as disclosed in awareness can be constructed as an entity formed by the sum of the impressions of isolated factors.

Again cogitation is a further limitation of fact in that it is a canalisation of consciousness so as to divest it of the crudeness of awareness. This illustrates that in limitation there is a gain in clarity, or definition, or intensity, but a loss of content.

For example, the factor red refers to fact as canalised by relationships of other factors to red, and the entity red is the factor red in its capacity as a relatum in the relationship of contrast, whereby it is contrasted with green or with sound or with the moon or with the multiplication table. Thus the factor red, essentially for its being, occasions the exhibition of a special aspect of fact, and the entity red is a further limitation of this aspect. Similarly the number three is nothing else than the aspect of fact as factors grouped in triplets. And the Tower of London is a particular aspect of the Universe in its relation to the banks of the Thames. Thus an entity is an abstraction from the concrete, which in its fullest sense means totality.

The point of this doctrine on which I want to insist is that any factor, by virtue of its status as a limitation within totality, necessarily refers to factors of totality other than itself. It is therefore impossible to find anything finite, that is to say, any entity for cogitation, which does not in its apprehension by consciousness disclose relationships to other entities, and thereby disclose some systematic structure of factors within fact. I call this quality of finitude, the significance of factors. This doctrine of significance necessitates that we admit that awareness requires a dual cognisance of entities. There can he awareness of a factor as signifying, and awareness of a factor as signified. In a sense this may be represented as an active or a passive cognisance of the entity. The entity is either cognised for its own sake, that is to say, actively, or it is cognised for the sake of other entities, that is to say, passively. If an entity is cognised actively, it is cognised for the sake of what it is in itself, for the sake of what it can make of the universe. I will call this sort of awareness of a factor, cognisance by adjective; since it is the character of the factor in itself which is then dominant in consciousness. Although in cognisance by adjective an entity is apprehended as a definite character in its relations to other entities, yet in a sense this type of cognisance marks a breakdown in relatedness. For the general relatedness of the character to other factors merely marks the fullness of its content, so that in effect the character is cognised for what it is in itself. Relationships to other factors occur in such cognisance only because the character is not itself apart from that ordering of fact.

When an entity is cognised passively, we are aware of it for the sake of some other factor. We are conscious passively of factor *A*, because factor *B* of which we are actively aware would not be what it is apart from its relatedness to *A*. Thus the individual character of *A* is in the background, and *A* becomes a vague something which is an element in a complex of systematic relatedness. The very nature of the relatedness may impose on *A* some character. But the character is gained through the relatedness and not the relatedness through the character. Accordingly *A* gains in consciousness the very minimum foothold for the relationship of contrast, and is thus the most shadowy of entities. I will call this sort of awareness of a factor, cognisance by relatedness. For example the knowledge of events inside another room is to be gained by their spatial and other relationships to events of which we have cognisance by adjective.

Thus cognisance of one factor by relatedness presupposes cognisance of other factors by adjective; and conversely, cognisance of one factor by adjective presupposes cognisance of other factors by relatedness.

It is possible to be aware of a factor both in cognisance by adjective and cognisance by relatedness. This will be termed 'full awareness' of the factor and is the usual form of awareness of factors within the area of clear apprehension when intrinsic characters and mutual relations are jointly apparent. 'Perception' will be the name given to the consciousness of a factor when to full awareness cogitation of it as an entity is also superadded.

But cogitation does not necessarily presuppose full awareness. For the contrast involved in cogitation may simply fall on the quality of the individualities of the factors, as when green as such is contrasted with red as such. In such a case merely awareness by adjective is presupposed. But the contrast may also fall on the specific relationships of each of the two factors to other factors, as when we contrast an event in the interior of the moon with another event in the interior of the earth. The spatio-temporal relationships of the two events are then contrasted; and it is from contrasts of this type that the two events gain their definite individuality as entities.

At this point in the discussion I will confine the scope of the remainder of my lecture strictly to the consideration of the relatedness of nature. This requires us to recognise another limitation within awareness which cuts across those already mentioned. I mean the limitation of awareness to sense-awareness. Nature is the system of factors apprehended in sense-awareness. But sense-awareness can only be defined negatively by enumerating what it is not.

Divest consciousness of its ideality, such as its logical, emotional, aesthetic and moral apprehensions, and what is left is sense-awareness. Thus sense-awareness is consciousness minus its apprehensions of ideality. It is not asserted that there is consciousness in fact divested of ideality, but that awareness of ideality and sense-awareness are two factors discernible in consciousness. The question as to whether either the one or the other, or both jointly may not be a factor necessary for consciousness is beyond the scope of the present discussion. The finiteness of individual consciousness means ignorance of what is there for knowledge. There is limitation of factors cognised by adjective, and equally there is limitation of factors cognised by relatedness. So it is perfectly possible to hold, as I do hold, that nature is significant of ideality, without being at all certain that there may not be some awareness of nature without awareness of ideality as signified by nature. It would have, I think, to be a feeble awareness. Perhaps it is more likely that ideality and nature are dim together in dim consciousness. It is unnecessary for us to endeavour to solve these doubts. My essential premise is that we are conscious of a certain definite assemblage of factors within fact and that this assemblage is what I call nature. Also I entirely agree that the factors of nature are also significant of factors which are not included in nature. But I propose to ignore this admitted preternatural significance of nature, and to analyse the general character of the relatedness of natural entities between themselves.

Nature usually presents itself to our imagination as being composed of all those entities which are to be found somewhere at some time. Sabre-toothed tigers are part of nature because we believe that somewhere and at some time sabre-toothed tigers were prowling. Thus an essential significance of a factor of nature is its reference to something that happened in time and space. I give the name 'event' to a spatio-temporal happening. An event does not in any way imply rapid change; the endurance of a block of marble is an event. Nature presents itself to us as essentially a becoming, and any limited portion of nature which preserves most completely such concreteness as attaches to nature itself is also a becoming and is what I call an event. By this I do not mean a bare portion of space-time. Such a concept is a further abstraction. I mean a part of the becomingness of nature, coloured with all the hues of its content.

Thus nature is a becomingness of events which are mutually significant so as to form a systematic structure. We express the character of the systematic structure of events in terms of space and time. Thus space and time are abstractions from this structure.

Let us now examine more particularly the significance of events in so far as it falls within nature. In this way we are treating nature as a closed system, and this I believe is the standpoint of natural science in the strict sense of the term.

But before embarking on the details of this investigation I should like to draw your attention to an objection, and a very serious objection, which is urged by opponents of the whole philosophic standpoint which I have been developing. You admit, it is said, that a factor is not itself apart from its relations to other factors. Accordingly to express any truth about one entity you must take into account its relations to all entities. But this is beyond you. Hence, since unfortunately a proposition must be either right or wrong or else unmeaning and a mere verbal jangle, the attainment of truth in any finite form is also beyond you.

Now I do not think that it is any answer to this argument to say that our propositions are only a little wrong, any more than it is a consolation to his friends to say that a man is only a little dead. The gist of the argument is that on our theory any ignorance is blank ignorance, because knowledge of any factor requires no ignorance. A philosophy of relatedness which cannot answer this argument must collapse, since we have got to admit ignorance.

Obviously if this argument is to be answered, I must guard and qualify some of the statements which have been made in the earlier portion of this lecture. I have put off the job until now, partly for the sake of simplicity, not to say too much at once, and also partly because the line of argument is most clearly illustrated in the case of nature, and indeed the application to nature is the only one in which for the purposes of this lecture we are interested. So I have waited until my discourse had led me to the introduction of nature.

The answer can only take one road, we must distinguish between the essential and the contingent relationships of a factor. The essential relationships of a factor are those relationships which are inherent in the peculiar individuality of the factor, so that apart from them the factor is not the special exhibition of finitude within fact which it is. They are the relationships which place the factor as an entity amid a definite system of entities. The significance of a factor is solely concerned with its essential relationships. The contingent relationships of a factor are those relationships between that factor and other factors which might be otherwise without change of the particular individuality of the factor. In other words, the factor would be what it is even if its contingent relationships were otherwise.

Thus awareness of a factor must include awareness of its essential relationships, and is compatible with ignorance of its contingent relationships.

It is evident that essential and contingent relationships correspond closely to internal and external relations. I hesitate to say how closely, since a different philosophic outlook radically affects all meanings.

We still have to explain how awareness of a factor can exclude ignorance of the relationships involved in its significance. For, on the face of it, this doctrine means that to perceive factor *A* we require also to perceive factors *B, C, D,* etc., which *A* signifies. In view of the possibilities of ignorance, such a doctrine appears to be extremely doubtful. This objection ignores the analysis of awareness into cognisance by adjective and cognisance by relatedness. In order to perceive *A* we do not require to be conscious of *B, C, D,* with cognisance by adjective. We only require cognisance by relatedness. In other words we must be conscious of *B, C, D,* ... as entities requisite for that relatedness to *A*, which is involved in *A*'s significance. But even this explanation asks for too much. It suggests that we must be conscious of *B, C, D*, ... as a definite numerical aggregate of entities signified by *A*. Now it is evident that no factor *A* makes us conscious of the individual entities of such an aggregate. Some necessary qualification of the doctrine of significance has been omitted. The missing principle is that any factor *A* has to be uniformly significant. Every entity involves that fact shall be patient of it. The patience of fact for *A* is the converse side of the significance of *A* within fact. This involves a canalisation within fact; and this means a systematic aggregate of factors each with the uniform impress of the patience of fact for *A*. *A* can be, because they are. Each such factor individually expresses the patience of fact for *A*.

Thus the knowledge required by the significance of *A* is simply this. In order to know *A* we must know how other factors express the patience of fact for *A*. We need not be aware of these other factors individually, but the awareness of *A* does require an awareness of their defining character. There is no such entity as mere *A* in isolation. *A* requires something other than itself, namely, factors expressing the patience of fact in respect to factor *A*.

Let us now apply to nature this doctrine of uniform significance. We commence by taking the case of the colour green. When we perceive green, it is not green in isolation, it is green somewhere at some time. The green may or may not have the relationship to some other object, such as a blade of grass. Such a relation would be contingent. But it is essential that we see it somewhere in space related to our eyes at a certain epoch of our bodily life. The detailed relationships of green to our bodily life and to the situations in which it is apparent to our vision are complex and variable and partake of the contingence which enables us to remain ignorant of them. But there can be no knowledge of green without apprehension of times and places. Green presupposes here and there, and now and then. In other words, green presupposes the passage of nature in the form of a structure of events. It may be merely green associated vaguely with the head, green all about me; but green is not green apart from its signification of events with structural coherence, which are factors expressing the patience of fact for green.

A blade of grass is an object of another type which signifies nature as a passage of events. In this respect it only differs from green in so far as its contingent relations to some definite events are perhaps sharper and capable of more precise determination.

The significance of events is more complex. In the first place they are mutually significant of each other. The uniform significance of events thus becomes the uniform spatio-temporal structure of events. In this respect we have to dissent from Einstein who assumes for this structure casual heterogeneity arising from contingent relations. Our consciousness also discloses to us this structure as uniformly stratified into durations which are complete nature during our specious presents. These stratifications exhibit the patience of fact for finite consciousness, but then they are in truth characters of nature and not illusions of consciousness.

Returning to the significance of events, we see that there is no such thing as an isolated event. Each event essentially signifies the whole structure. But furthermore, there is no such entity as a bare event. Each event also signifies objects, other than events which are in essential relation to it. In other words the passage of an event exhibits objects which do not pass. I have termed the natural factors which are not events but are implicated in events 'objects,' and awareness of an object is what I have termed recognition. Thus green is an object and so is a blade of grass, and awareness of green or of a blade of grass is recognition. Thus an event signifies objects in mutual relations. The particular objects and their particular relations belong to the sphere of contingence; but the event is essentially a 'field,' in the sense that without related objects there can be no event. On the other hand related objects signify events, and without such events there are no such objects.

The celebrated two-termed relationship of universals to the concrete particulars which they qualify is merely a particular example of the general doctrine of significance and patience. The universals are significant of their particulars, and the particulars are factors exhibiting the patience of fact for those universals.

But in the apparent world, that is to say, in the world of nature disclosed by sense-awareness, no example of the simple two-termed relationship of a universal signifying its particular is to be found. Green appears to an observer in a situation distinct from that of the observer, but simultaneous with it. Thus there is essential reference to three simultaneous events, the event which is the bodily life of the observer, called the percipient event, and the event which is the so-called situation of the green at the time of observation, and to the time of observation which is nothing else than the whole of nature at that time. Under the obsession of the logical theory of universals and concrete particulars the percipient event was suppressed, and the relation of green to its situation represented as universal qualifying particular. It was then noted that this relation only holds for the particular observer, and that furthermore account must be taken of contingent circumstances such as the transmission of something, which is not the colour green, from an antecedent situation to the percipient event.

This process, of first presupposing a two-termed relation and then finding that it is not true, has led to the bifurcation which places green in the observer's mind, qualifying a particular also in the observer's mind; while the whole mental process has some undetermined relation to another system of entities variously described either as an independent physical universe in some causal relation to mind or as a conceptual model.

I have argued elsewhere in detail that this result is untenable. Here I will only remark that if we incline to adopt the physical universe, we can find no shred of evidence for it, since everything apparent for consciousness has been accounted for as being in the observer's mind; while, if we turn to the conceptual model, it is also the model for the same consciousness. Accordingly whichever choice we make there will be no shred of evidence for anything other than the play of that consciousness at one moment of self-realisation. For recollection and anticipation are merely the play of immediate consciousness. Thus on either alternative, solipsism only is left and very little of that.

Meanwhile the whole difficulty has arisen from the initial error of forcing the complex relations between green and the structure of events into the inadequate form of a two-termed relation.

Yet after all the search for universals to qualify events in the simple two-termed manner does represent a justifiable demand. We want to know what any particular event *A* is in itself apart from its reference to other events. By this I mean, we want to determine how *A* can enter into a two-termed relation of contrast with any other factor *X* without having necessarily to enlarge the relationship by including other events *B, C, D,* by way of determining *A*. For example, the colour green is in itself different from red, and we do not have to specify green or red by their diverse relationships amid events in order to appreciate their contrast. Now we want to do much the same thing for events, so as to feel that an event has a character of its own. We have seen that the immediate objects of the apparent world such as colours do not satisfy the requisite conditions since their reference to events involves the relations of the percipient event to the so-called situation. I call such objects of immediate appearance, sense-objects. Colours, sounds, smells, touches, pushes, bodily feelings, are sense-objects. But after all, the way we do connect these sense-objects, as I call them, with their situations shows that awareness of an event carries with it apprehension of that event as patient of a character qualifying it individually. In fact every event signifies a character for itself alone, but what exactly that character may be lies within the sphere of contingency and is not disclosed in our immediate consciousness of the apparent world. I will call such a character an adjective of its event. An adjective marks a breakdown in relativity by the very simplicity of the two-termed relation it involves. The discovery of these missing adjectives is the task of natural science. The primary aim of science is to contract the sphere of contingency by discovering adjectives of events such that the history of the apparent world in the future shall be the outcome of the apparent world in the past. There obviously is some such dependence, and it is the purpose of science to express this dependence in terms of adjectives qualifying events. In order to understand this procedure of science, there are three concepts which we must understand. They are

(i) The structure of the four dimensional continuum,
(ii) Pervasive adjectives and adjectival particles,
(iii) The atomic field of an adjectival particle.

I will conclude this lecture by considering them in order.

(i) *The structure of the continuum of events*

This structure is four-dimensional, so that any event is a four-dimensional hyper-volume in which time is the fourth dimension. But we should not conceive an event as space and time, but as a unit from which space and time are abstracts.

An event with all its dimensions ideally restricted is called an 'event-particle,' and an event with only one dimension of finite extension is called a 'route' or 'path.' I will not in this lecture discuss the meaning of this ideal restriction. I have investigated it elsewhere under the name of 'extensive abstraction.'

The structure is uniform because of the necessity for knowledge that there be a system of uniform relatedness, in terms of which the contingent relations of natural factors can be expressed. Otherwise we can know nothing until we know everything. If *P* be any event-particle, a moment through P is a system of event-particles representing all nature instantaneously contemporaneous with *P*. According to the classical view of time there can be only one such moment. According to the modern view there can be an indefinite number of alternative moments through *P*, each corresponding to a different meaning for time and space. A moment is an instantaneous three-dimensional section of nature and is the entity indicated when we speak of a moment of time.

The aggregate of event-particles lying on moments through *P* will be called the region co-present with *P*. The remainder of the four-dimensional continuum is divided by the co-present region into two regions, one being *P*'s past and the other being P's future. The three-dimensional boundary between *P*'s past and *P*'s co-present region is *P*'s causal past, and the corresponding boundary between *P*'s future and *P*'s co-present region is *P*'s causal future. The remaining portion of *P*'s future is *P*'s kinematic future.

A route lying entirely in one moment is called a spatial route, and a route which lies entirely in the past and future of each one of its event-particles is called a historical[1] route.

(ii) *Pervasive adjectives and adjectival particles*

We gain great simplicity of explanation, without loss of any essential considerations by confining our consideration of events to routes. These routes are of course not true events, but merely ideal limits with only one dimensional extension remaining.

A factor will be said to be an adjective pervading a route when it is an adjective of every stretch of the route. Such a factor will be called a pervasive adjective, or uniform object. I think—without being very certain—that true pervasive adjectives are only to be found qualifying historical routes; but that pervasive pseudo-adjectives also qualify spatial routes. The essential difference between time and space finds its illustration in the difference between these two different types of route.

As an illustration of pervasive adjectives, consider a mass-particle *m*. The enduring existence of this particle marks out a historical route amid the structure of events. In fact the mass-particle is merely a pervasive adjective of that route, since it is an adjective qualifying in the same sense every stretch of that route. But here a further explanation is necessary. The mass-particle as a pervasive adjective is a universal and has lost its concrete individuality.

Another mass-particle of the same mass pervading another historical route is the same pervasive adjective also qualifying every stretch of that other route. It follows that the separate concrete indi-

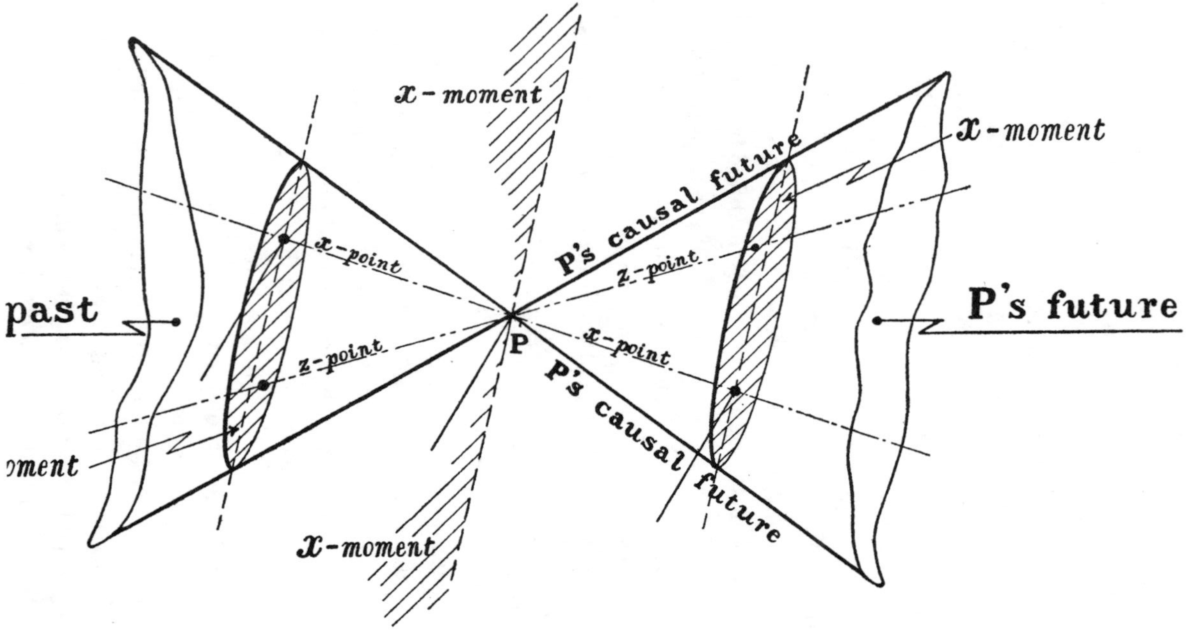
x-moment
x-moment
P's causal future
x-point
z-point
past
P
P's future
z-point
x-point
P's causal future
oment
x-moment

vidualities of the two mass-particles arise from the separate individualities of their two historical routes. Thus a concrete mass-particle is the fusion of a pervasive adjective with the individuality of a historical route. We say that a mass-particle is situated at each event-particle of its historical route. I will call a pervasive adjective as qualifying a particular historical route an 'adjectival particle.' The principle underlying the conception of an adjectival particle is that the individual embodiment of character always involves process and that this process is here represented by the historical route.

Spatial routes cannot be pervaded by mass-particles. Thus if a mass-particle of the same mass be situated at each event-particle of a spatial route, that route is not pervaded by the one adjective which is the same universal for each of the concrete mass-particles. In fact a stretch of the spatial route is qualified by quite a different adjective, which represents the sum of the masses situated at the event-particles of the stretch. Accordingly spatial routes and historical routes function quite differently in respect to the adjective 'mass *m*,' and thus illustrate the difference between the spread of space and the lapse of time.

There are however pseudo-adjectives which do pervade spatial routes. Consider a sense-object, such as the colour red. It is not a true adjective of its situation, since there is always a necessary reference to a percipient event. But for the one observer conscious of the natural relations of that percipient event, who is presupposing this reference to his bodily life as a condition for appearance, the colour red is an adjective of its situation. But any part of its situation is also red, neither more nor less so than the whole enduring patch of red. Thus red pervades its situation. However I have already argued at length that sense-objects are not true adjectives. They simulate adjectives for an observer who in his intellectual analysis of the circumstances forgets to mention himself. Accordingly they may be called pervasive pseudo-adjectives.

The common material objects of perception, such as chairs, stones, planets, trees, etc., are adjectival bodies pervading the historical events which they qualify. In so expressing myself, I have gone beyond the ideal simplicity of a route, and the terms 'pervasion' and 'historical event' require, strictly speaking, a more elaborate explanation than I have yet given. In this lecture however it is unnecessary to undertake the task, and I need only refer to my *Principles of Natural Knowledge* where the requisite definitions are given in connection with uniform objects.

(iii) *The atomic field of an adjectival particle*

Science has been driven to have recourse to more precisely delimited adjectival objects than these adjectival objects of perception. The standard types of such objects are mass-particles and electrons; and we will fall back on our ideal simplicities by conceiving them as adjectival particles defined, as above, for the ideal simplification of historical routes.

Now the essence of an adjectival object, whether it be the unprecise object of perception or the more precise object of science, is that it reduces the contingency of nature. It is an adjective of events which to some extent conditions the possibilities of apparent sense-objects.

It must be admitted that it is itself a contingent adjective. But owing to the simplicity of the relation of an adjective to its qualified substance, it involves a simpler contingency than the contingency of the complex relationships of sense-objects. In other words we are limiting contingency by the fixed conditions which are the laws of nature.

It is evident therefore that a scientific object must qualify future events. For otherwise the future contingency is unaffected by it. In this a scientific object differs decisively from a sense-object viewed as a pseudo-adjective. A sense-object qualifies events in the present. It is confined to a spatial region with the minimum of historicity requisite for the duration of the present. Whereas the scientific object qualifies a region extending from the present into the future. Thus the seemingly contingent play of the senses is controlled by the conditions introduced by its dependence upon the qualification of events introduced by scientific objects. A scientific object qualifies the future in two ways, (*a*) by its permanence and (*b*) by its field.

Let us take the permanence first. The permanence of an adjectival particle lets us know that there will be some historical route pervaded by that particle. It does not in itself tell us more than that some pervaded route will stretch into the future from the situation in the present. The permanence of the unique particle is nothing else than the continuity of the unique historical route, and its pervasion by the adjectival particle.

The further laws of physical science represent the further conditions which determine, or partially determine, the particular historical routes pervaded by these adjectival particles. The most simple expression of such a law consists in associating an atomic field with each adjectival particle as situated in each one of the event-particles of its pervaded route. Again this association of the field represents another eruption of contingency, but also again this contingency is of a simple defined type. The field of an adjectival particle *m* at a situation *P* is a limited region stretching from *P* into *P*'*s* futurity. This region is qualified by an adjective dependent upon *m* and *P* only. For this simple type of law, the only limited region which can satisfy this demand is the three-dimensional boundary region between *P*'s co-present and *P*'s kinematic future. I have called this region *P*'s causal future. Accordingly the field of *m* at *P* must be *P*'s causal future. Expressing this statement in terms of one consistent meaning for time and its associated permanent space, we first note that *P* consists of a point S_P at a time t_P, and *m* situated in *P* means *m* at the point *p* at the time *p*. The causal future of *P* means those points S_B, at times t_B [i.e. those event-particles such as *B*], reached by a physical character due to *m*, and starting from S_P at time t_P and arriving at S_B at time t_B, and travelling with the critical maximum velocity *c*.

Experiment shows that this critical maximum velocity is a near approximation to the velocity of light *in vacuo*, but its definition in no way depends upon any reference to light. Thus the adjectival character of the field of *m* at *P* consists in the correlated physical characters of the different event-particles of the field. The whole conception is practically the familiar one of tubes of force, with one exception. A tube of force is conceived statically as a simultaneous character stretching through space. This statical conception destroys the true individuality of a tube by piecing together fragments of different tubes. As we pass along a tube radiating from S_P we keep to the same tube by allowing for the lapse of time required by the velocity *c*.

The peculiar correlation of adjectives attaching to the various event-particles of the field of *m* at *P* will depend upon the particular contingent law which science conjectures to be the true expression of *m*'s physical status.

There are, also, less simple laws of nature for which the influence of the contingent configurations of other adjectival particles will be essential factors. Such laws will in general involve the deflection of the field of *m* at *P* from *P*'s causal future into *P*'s kinematic future. The region will be dependent upon the fields of the other relevant adjectival particles. It is evident that with such laws we are rapidly drifting towards the difficulty of having to know everything before knowing anything.

I will call such fields 'obstructed fields.' Differential equations help us here. But even their aid would be unavailing unless we could approximate from the first assumption of unobstructed fields for the adjectival particles producing the obstruction. In this way the influence of gravitation upon the electromagnetic field can be calculated and *vice versa.*

This account of the status of scientific objects completely changes the status of the ether from that presumed in nineteenth century science. In the classical doctrine the ether is the shy agent behind the veil: in the account given here the ether is exactly the apparent world, neither more nor less. The apparent world discloses itself to us as the ingression of sense-objects amid events. In this statement the term 'ingression' is used for the complex relationship of those abstract elements of the world, such as sense-objects, which are devoid of becomingness and extension, to those other more concrete elements (events) which retain becomingness and extension. But a bare event is a mere abstraction. Events are disclosed as involved in this relationship of ingression. This disclosure is our perceptual vision of the apparent world. We now ask on behalf of science whether we cannot simplify the regulative principles discerned in this apparent world by treating events as something more than relata in the relationship of ingression. Cannot we discern true Aristotelian qualities as attaching to the events? Is not each event something in itself, apart from its status as a mere relatum in the relationship of ingression? The apparent world itself gives an answer, partially in the affirmative. Chairs, tables, and perceptual objects generally, have lost the complexity of ingression, and appear as the required Aristotelian adjectives of some events. Their appearance involves that borderline where sense-awareness is fusing with thought. It is difficult to make any account of them precise. In fact, for the purpose of science they suffer from incurable vagueness. But they mark the focal centres to be used as the radiating centres for an exact account of true Aristotelian adjectives without any of those qualifications here referred to as 'vagueness.' The events of the apparent world as thus qualified by the exact adjectives of science are what we call the 'ether.' Accordingly in my previous work, *The Principles of Natural Knowledge,* I have phrased it in this way, that the older 'ether of stuff' is here supplanted by an 'ether of events.'

This line of thought, supplanting 'stuff' by 'events,' and conceiving events as involving process and extension and contingent qualities and as primarily relata in the relationship of ingression, is a recurrence to Descartes' views—with a difference. Descartes, like the rest of the world at that time, completely dissociated space and time. He assigned extension to space, and process to time. It is true that time involves extension of some sort, but that does not seem to have coloured his philosophy. Now according to Descartes 'extension' is an abstract from the more concrete concept of 'stuff.' He, like the rest of the world, considers stuff as being separable from the concept of 'process,' so that stuff fully realises itself *at* an instant, without duration. Space is thus a property of stuff, and accordingly follows stuff in being essentially dissociated from time. He therefore deduces that space is an essential timeless plenum. It is merely an abstract from the concrete world of appearance *at an instant.* If there be no stuff to appear, there can be no space.

Now re-write this Cartesian account of space, substituting 'events' (which retain 'process') for 'stuff' (which has lost 'process'). You then return to my account of space-time, as an abstract from events which are the ultimate repositories of the varied individualities in nature. But space as pure extension, dissociated from process, and time as pure serial process, are correlative abstractions which can be made in different ways, each way representing a real property of nature. In this manner the alternative spaces and the alternative times, which have already been mentioned, are seen to be justifiable conceptions, according to the account of the immediate deliverances of awareness here given, provided that our experience can be thereby explained.

Mere deductive logic, whether you clothe it in mathematical symbols and phraseology or whether you enlarge its scope into a more general symbolic technique, can never take the place of clear relevant initial concepts of the meaning of your symbols, and among symbols I include words. If you are dealing with nature, your meanings must directly relate to the immediate facts of observation. We have to analyse first the most general characteristics of things observed, and then the more casual contingent occurrences. There can be no true physical science which looks first to mathematics for the provision of a conceptual model. Such a procedure is to repeat the errors of the logicians of the middle-ages.

CHAPTER III

EQUALITY

THE criticism of the meanings of simple obvious statements assumes especial importance when any large reorganisation of current ideas is in progress. The upheaval produced by the Einstein doctrine of relativity is a case in point. It demands a careful scrutiny of the fundamental ideas of physical science in general and of mathematical physics in particular. I propose therefore in this lecture to take one of the simplest mathematical notions which we all come across when we start mathematics in our early school life and to ask what it means.

The example I have chosen is the notion of 'equality.' There is hardly a page or a paragraph of any mathematical book which does not employ this idea. It appears in geometry in the more specialised form of congruence.

If I am not mistaken, clear notions on equality are of decisive importance for the sound reconstruction of mathematical physics. Congruence is a more special term than equality, being confined to mean the quantitative equality of geometrical elements. Equality is also closely allied to the idea of quantity; but here again I think that equality touches the more general ideas. The consideration of quantity necessarily introduces that of measurement. In fact the scope of a discussion on quantity may be defined by the question, How is measurement possible? Lastly, equality has an obvious affinity with identity. Some philosophers in considering the foundations of mathematics would draw no distinction between the two. In certain usages of equality this may be the case. But it cannot be the whole truth. For if it were, the greater part of mathematics would consist of a reiteration of the tautologous statement that a thing is itself. We are interested in equality because diversity has crept in.

In fact a discussion of equality embraces in its scope congruence, quantity, measurement, identity and diversity. The importance of equality was discovered by the Greeks. We all know Euclid's axiom, 'Things that are equal to the same thing are also equal to one another' (*τὰ τῷ αὐτῷ ἴσα καὶ ἀλλήλοις ἐστὶν ἴσαῷ*). This axiom deserves its fame, in that it is one of the first efforts to clarify thought by an accurate statement of premises habitually assumed. It is the most conspicuous example of the decisive trend of Greek thought towards rigid accuracy in detailed expression, to which we owe our modern philosophy, our modern science, and the creeds of the Christian Church. But grateful as we are to the Greeks for this axiom and for the whole state of mind which it indicates, we cannot withdraw it from philosophic scrutiny. The whole import of the axiom depends on the meaning of the word *ἴσος*, equal. What do we mean when we say that one thing is equal to another? Suppose we explain by stating that 'equal' means 'equal in magnitude,' that is to say, the things are quantities of the same magnitude. But what is a quantity? If we define it as having the property of being measurable in terms of a unit, we are thrown back upon the equality of different examples of the same unit. It is evident that we are in danger of soothing ourselves with a vicious circle whereby equality is explained by reference to quantity and quantity by reference to equality.

Let us first drop the special notion of quantitative equality and consider the most general significance of that notion. The relation of equality denotes a possible diversity of things related but an identity of character qualifying them. It is convenient for technical facility in the arrangement of deductive trains of reasoning to allow that a thing is equal to itself, so that equality includes identity as a special case. But this is a mere matter of arbitrary definition.

The important use of equality is when there is diversity of things related and identity of character. This identity of character must not be mere identity of the complete characters. For in that case, by the principle of the identity of indiscernables, the equal things would be necessarily identical.

Accordingly when we write

$$A = B$$

we are referring implicitly to some character and asserting that *A* and *B* both possess it. The assertion of equality is therefore generally couched in a highly elliptical form since the expression of the character in question is often omitted. This is a source of most of the confused thinking which haunts discussion on this subject. Let us remedy our notation so as to rid it of its misleading ellipticity. Let $(c_1, c_2, \ldots, c_n)$ denote a class of characters $c_1, c_2, \ldots, c_n$, such as colour for example.

Then we write

$$A = B \rightarrow (c_1, c_2, \ldots, c_n)$$

to mean that *A* and *B* both possess the same character out of the set $(c_1, c_2, \ldots, c_n)$; and we write

$$A \neq B \rightarrow (c_1, c_2, \ldots, c_n)$$

to mean that different characters out of the set apply to *A* and *B* respectively. Our notation still has the defect of implying that the class of characters is a finite or at least an enumerable class. Let us therefore take γ to represent this class, so that

$$A = B \rightarrow \gamma$$

means that the same member of the class γ qualifies both *A* and *B;* and

$$A \neq B \rightarrow \gamma$$

means that one member of γ qualifies *A* and that another member of γ qualifies *B*. I will call γ the 'qualifying class.'

It is now evident that

$$A = B \rightarrow \gamma$$

and $$B = C \to \gamma$$

implies that $$A = C \to \gamma.$$

This is evidently a general rendering of Euclid's first axiom.

But we are not yet at the end of our discussion. In the first place, we cannot yet prove that

$$A = B \to \gamma$$

and $$A \neq B \to \gamma$$

are incompatible with each other. For we have not yet excluded the case that more than one character of the set γ may attach either to A or to B or to both. For example if c_1 and c_2 are members of γ, both attaching to A, but only c_1 attaching to B, then both

$$A = B \to \gamma$$

and $$A \neq B \to \gamma.$$

Accordingly we must re-define the meaning of our symbols by introducing the additional limitation that

$$A = B \to \gamma$$

and $$A \neq B \to \gamma$$

both mean that A and B each possess one and only one character of the class γ. It is well to note that the two propositions represented by these symbolic statements are only contraries to each other. For though they cannot both be true, they will both be false if either A or B does not possess any character out of the qualifying class γ. For example if A does not possess any such quality or if it possesses two such qualities, then

$$A = A \to \gamma$$

and $$A = A \to \gamma$$

are both false. This example also illustrates the sharp distinction between equality and mere identity.

In this most general sense of equality, the notion of 'matching' in the sense in which colours match, might with advantage replace equality, so that we should interpret

$$A = B \to \gamma \text{ and } A \neq B \to \gamma$$

as meaning respectively

'*A* matches *B* in respect to the qualities γ'

and

'*A* does not match *B* in respect to the qualities γ.' This verbal statement in its common meaning presupposes our three conditions:

(i) that *A* and *B* each possess one of the qualities γ,

(ii) that neither *A* nor *B* possesses more than one such quality,

(iii) that *A* and *B* possess the same one of the qualities γ, and (in the second case) that *A* and *B* do not possess the same one of the qualities.

The set of entities such as *A* and *B* possessing one and only one of the qualities of the class γ will be said to form the 'qualified class for γ,' and we have already named γ the 'qualifying class.'

Congruence. Congruence is a subspecies of the general type of the equality relation. Let us start with the simplest example and consider a one-dimensional space. The points of this space are terms interconnected by a relation which arranges them in serial order with the ordinary continuity of the Dedekindian type. The points may be connected by other relations which sort them out in other ways; but when we say that they form a one-dimensional space, we are thinking of one definite relation which produces the continuous serial order, both ways infinite.

Now in the particularising of the equality relation so as to produce a congruence relation for this space, we first demand that, if γ be the qualifying class, the class qualified by γ must be composed of all the finite stretches of the space. Thus the terms *A*, *B*, etc. in the previous explanation of equality are now stretches of the serial space, and every finite stretch belongs to the qualified class. It will be convenient to confine attention to those stretches which include their two end-points. Let two stretches which do not overlap, except that they have one end-point in common, be called adjoined stretches, or stretches adjoined at that end-point.

Now the conditions which have to be fulfilled in order that this type of equality may reckon as a congruence are:

(i) If A be any stretch and p any point, there are two stretches P_1 and P_2 adjoined at p, such that

$$A = P_1 \rightarrow \gamma$$

and

$$A = P_2 \rightarrow \gamma.$$

In other words, from a given point p stretches of an assigned length can be measured in either direction.

(ii) If P and Q are two stretches, and P contains Q, then

$$P \neq Q \rightarrow \gamma.$$

In other words, the whole is unequal to its part.

(iii) If P and Q be two stretches, and P be composed of the adjoint stretches P_1 and P_2, and Q of the adjoint stretches Q_1 and Q_2, and furthermore if

$$P_1 = Q_1 \rightarrow \gamma$$

and

$$P_2 = Q_2 \rightarrow \gamma,$$

then

$$P = Q \rightarrow \gamma.$$

In other words, if equals be added to equals the wholes are equal.

(iv) If the first clause of the hypothesis of (iii) hold, and furthermore if

$$P = Q \rightarrow \gamma$$

and

$$P_1 = Q_1 \rightarrow \gamma,$$

then

$$P_2 = Q_2 \rightarrow \gamma.$$

In other words, if equals be taken from equals the remainders are equal.

(v) The axiom that the whole is greater than its part suffers from the difficulty that we have not defined what we mean by 'greater than.' Our condition (ii) states that the whole is unequal to its part. But the idea of 'greater than' really follows from the condition which we wish to express. I think that the missing condition is best stated thus:

Let A and B be two stretches of which one contains the other, so that either A contains B or B contains A, and let H and K be two other stretches with the same property in regard to each other.

Also let

$$A = H \rightarrow \gamma,$$

$$B = K \rightarrow \gamma.$$

Then if H contains K, it also follows that A contains B. The point of this condition is that we exclude the crosswise equality in which A is congruent to a part of H and H to a part of A.

Then the idea of any stretch P being greater than any stretch Q must be defined to mean that there is a stretch H containing a part K such that

$$P = H \rightarrow \gamma,$$
$$Q = K \rightarrow \gamma.$$

Thus the verbal form, the whole is greater than its part, becomes a mere tautology. The true point being first our condition (ii) that the whole is unequal to any of its parts, and our condition (v) which excludes the crosswise equality of wholes to parts.

The theory of numerical measurement depends upon three additional conditions which can be conveniently preceded by some definitions. Let a sequence of n successively adjoined stretches $A_1, A_2, \ldots, A_n$, which is such that

$$A_p = A_q \rightarrow \gamma, \; [p, q = 1, 2, \ldots, n]$$

be called a 'stretch sequence for γ.' Let each individual stretch of the sequence be called a 'component stretch' of the sequence, and let the stretch which is composed of all the stretches of the sequence be called the 'resultant stretch' of the sequence.

Furthermore if c be the member of γ which characterises each component stretch of the sequence of n stretches, let nc be the symbol for the member of γ which characterises the resultant stretch of the sequence.

Also if c' be an alternative symbol for nc, let $\frac{1}{n}c'$ be an alternative symbol for c.

The three conditions are:

(vi) If A be any stretch and n be any integer, then a stretch sequence for γ can be found composed of n members such that A is its resultant.

(vii) If A and B be any two coterminous stretches, and A be part of B, then we can find an integer n such that there exists a stretch sequence for γ of n terms such that A is its first term and B is part of the resultant of the sequence.

(viii) If A be any stretch and n any integer, then A is a member in any assigned ordinal position of two stretch sequences for γ of n terms, the two sequences running in opposed directions.

The condition (vii) is the axiom of Archimedes.

It is evident that we may conceive γ as the class of magnitudes and the stretches as the class of concrete quantities. The difference between a magnitude and a concrete quantity is the difference between the length, called a yard, and the particular concrete instance which is in the custody of the Warden of the Standards.

It is not necessary to plunge further into the exact analysis of the theory of extensive quantity. The discussion has been carried far enough to make it evident that the qualifying class γ, which is the class of magnitudes, is simply a class of qualities which happen to be sorted out among the qualified class (which in the above example was a class of stretches) in such a way that, when one member of γ has been taken as the standard of reference, the unit, all the other members of γ can be described in terms of it by means of real numbers. But a quality which belongs to the set γ is in itself in no way otherwise distinguished from any other quality of things. Quantity arises from a distribution of qualities which in a certain definite way has regard to the peculiar fact that in certain cases two extended spatio-temporal elements together form a third such element. In fact the 'qualifying' qualities are distributed among extended things with a certain regard to their property of extension. Also it is evident that two stretches A and B which are equal for one qualifying class y may be unequal for another qualifying class γ'.

If we apply this doctrine to the classical theory of space and time, we find, following Sophus Lie's analysis, that there are an indefinite number of qualifying classes γ, γ', γ'', etc., which for the case of three-dimensional space generate relations of congruence among spatial elements, and that each such set of congruence relations is inconsistent with any other such set.

For the case of time the opposite trouble arises. Time in itself, according to the classical theory, presents us with no qualifying class at all on which a theory of congruence can be founded.

This breakdown of the uniqueness of congruence for space and of its very existence for time is to be contrasted with the fact that mankind does in truth agree on a congruence system for space and on a congruence system for time which are founded on the direct evidence of its senses. We ask, why this pathetic trust in the yard-measure and the clock? The truth is that we have observed something which the classical theory does not explain.

It is important to understand exactly where the difficulty lies. It is often wrongly conceived as depending on the inexactness of all measurements in regard to very small quantities. According to our methods of observation we may be correct to a hundredth, or a thousandth, or a millionth of an inch. But there is always a margin left over within which we cannot measure. However this character of inexactness is not the difficulty in question.

Let us suppose that our measurements can be ideally exact; it will be still the case that if one man uses one qualifying class γ and the other man uses another qualifying class δ, and if they both admit the standard yard kept in. the exchequer chambers to be their unit of measurement, they will disagree as to what other distances places should be judged to be equal to that standard distance in the exchequer chambers. Nor need their disagreement be of a negligible character. For example, the man who uses the qualifying class γ might be in agreement with the rest of us, who are also using γ , and the other man who uses δ might also be a well-trained accurate observer. But in his measurement the distance from York to Edinburgh might come out at exactly one yard.

But no one, who is not otherwise known to be a lunatic, is apt to make such a foolish mistake.

The conclusion is that when we cease to think of mere abstract mathematics and proceed to measure in the realm of nature, we choose our qualifying class γ for some reason in addition to the mere fact that the various characters included in γ are sorted among stretches so as to satisfy the conditions for congruence which I have jotted down above.

When we say that two stretches match in respect to length, what do we mean? Furthermore we have got to include time. When two lapses of time match in respect to duration, what do we mean? We have seen that measurement presupposes matching, so it is of no use to hope to explain matching by measurement.

We have got to dismiss from our minds all considerations of number and measurement and quantity, and simply concentrate attention on what we mean by matching in length.

It is an entirely different and subsequent consideration as to whether length in this sense of the term is a class of qualities which is sorted out to stretches in accordance with the congruence conditions.

Our physical space therefore must already have a structure and the matching must refer to some qualifying class of qualities inherent in this structure. The only possible structure is that of planes and straight lines, such that stretches of straight lines can be conceived as composed of points arranged in order.

An additional factor of structure can be that of ordinary Euclidean parallelism. By this I mean that through any point outside a plane there is one and only one plane which does not intersect a given plane. You will observe that I have had to adopt what is termed Playfair's axiom for the definition of parallels. It is the only one which does not introduce some presupposition of congruence, either of length or angles. I draw your attention to the absolute necessity of defining our structure without the presupposition of congruence. If we fail in this respect our argument will be involved in a vicious circle.

With this definition of parallels it is now very easy to get some way in the explanation of what we mean by stretches matching in length. For since our structure includes parallels, it also includes parallelograms. Accordingly we can agree that the opposite sides of parallelograms match in length. It is then easy enough to show that we have a complete system of congruence for any one system of parallel stretches in space. This means that if there are any two stretches either on the same straight line or on parallel straight lines, we have a definitely determined numerical ratio of the length of one to the length of the other.

But we cannot go further and compare the lengths of two stretches which are not parallel, unless we introduce some additional principle for the matching of lengths.

We can find this additional principle provided that we can define a right-angle without any appeal to the idea of congruence or equality. For let us anticipate such a definition independent of congruence.

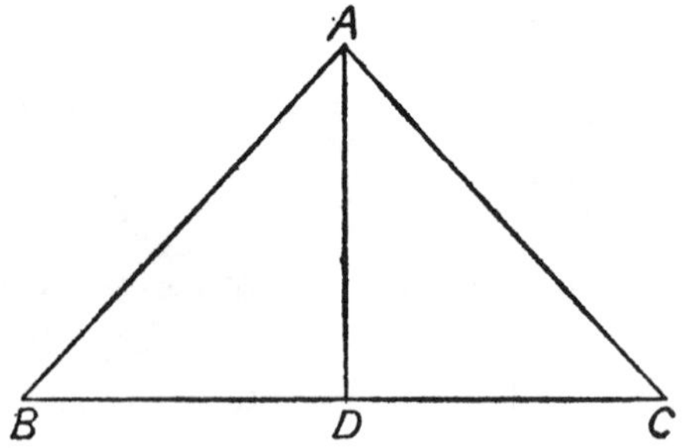

Let *D* he the midpoint of the stretch *BC*, and draw *DA* perpendicular to *BC*. Then our additional principle of matching shall be that *AB* is equal to *AC*. In this way we can compare the lengths of stretches which are not parallel, and the whole theory of congruence in space is established.

But as yet we have not gone any way towards finding any theory for the congruence of lapses of time. Accordingly if we are to explain how it is that in our observation of nature we all agree in our systems of space and time congruence, we have to explain what we mean by planes, and by right-angles, and how we should match lapses of time. We can omit straight lines from this catalogue, since they can be defined as the intersections of planes. We shall however have to explain how the points on straight lines come to be arranged in order.

When we are conscious of nature, what is it that we really observe? The obvious answer is that we perceive various material bodies, such as chairs, bricks, trees. We can touch them, see them and hear them. As I write I can hear the birds singing in a Berkshire garden in early spring.

In conformity with this answer, it is now fashionable and indeed almost universal to say that our notions of space merely arise from our endeavours to express the relations of these bodies to each other. I am sorry to appear pigheaded; but, though I am nearly in a minority of one, I believe this answer to be entirely wrong. I will explain my reasons.

Are these material bodies really the ultimate data of perception, incapable of further analysis?

If they are, I at once surrender. But I submit that plainly they have not this ultimate character. My allusion to the birds singing was made not because I felt poetical, but to warn you that we were being led into a difficulty. What I immediately heard was the song. The birds only enter perception as a correlation of more ultimate immediate data of perception, among which for my consciousness their song is dominant.

Material bodies only enter my consciousness as a representation of a certain coherence of the sense-objects such as colours, sounds and touches. But these sense-objects at once proclaim themselves to be adjectives (pseudo-adjectives, according to the previous chapter) of events. It is not mere red that we see, but a red patch in a definite place enduring through a definite time. The red is an adjective of the red time and place. Thus nature appears to us as the continuous passage of instantaneous three-dimensional spatial spreads, the temporal passage adding a fourth dimension. Thus nature is stratified by time. In fact passage in time is of the essence of nature, and a body is merely the coherence of adjectives qualifying the same route through the four-dimensional space-time of events.

But as the result of modern observations we have to admit that there are an indefinite number of such modes of time stratification.

However, this admission at once yields an explanation of the meaning of the instantaneous spatial extension of nature. For it explains this extension as merely the exhibition of the different ways in which simultaneous occurrences function in regard to other time-systems.

I mean that occurrences which are simultaneous for one time-system appear as spread out in three dimensions because they function diversely for other time-systems. The extended space of one time-system is merely the expression of properties of other time-systems.

According to this doctrine, a moment of time is nothing else than an instantaneous spread of nature. Thus let t_1, t_2, t_3 be three moments of time according to one time-system, and let T_1, T_2, T_3 be three moments of time according to another time-system. The intersections of pairs of moments in diverse time-systems are planes in each instantaneous three-dimensional space. In the dia-

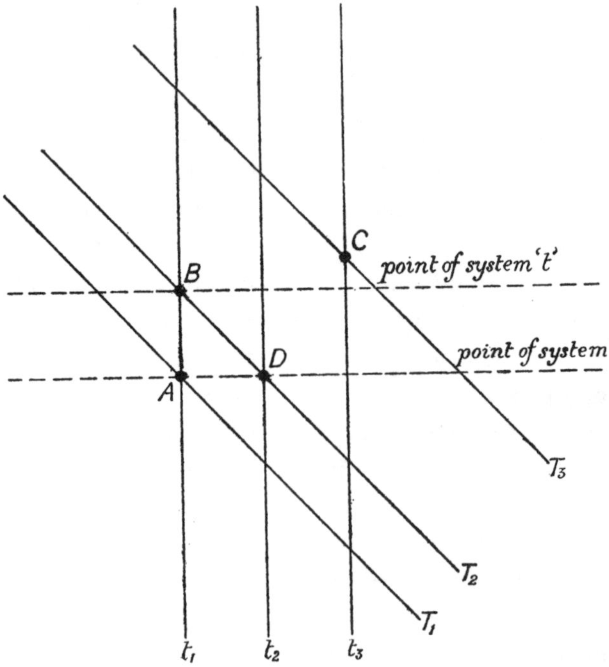

gram each continuous line accordingly symbolises a three-dimensional space; and the intersections of continuous lines, such as *A* or *B* or *C*, symbolize planes. Thus t_1 and T_1 are each a three-dimensional space, and *A* is a plane in either space.

Parallelism is the reflection into an instantaneous space of one time-system of the property of moments of some other system. Thus *A* and *B* are parallel planes in t_1, since T_1 and T_2 are moments of the same system which is not the system to which t_1 belongs.

But when we talk of space we are not usually thinking of the instantaneous fact of immediate perception. We are thinking of an enduring scheme of extension within which all these instantaneous facts are fitted. It follows that we ought to be able to find a meaning for the idea of a permanent space in connection with each time-system.

This conception must arise from our immediate observations of motion and rest. Both rest and motion have no meaning in connection with one mere instantaneous space. In such a space everything is where it is and there is an instantaneous end to it; to be succeeded by another instantaneous space. But motion and rest at once warn us that our perception involves something more.

The instantaneous moment is merely an ideal limit of perception. Have you ever endeavoured to capture the instantaneous present? It eludes you, because in truth there is no such entity among the crude facts of our experience. Our present experience is an enduring fact within which we discriminate a passage of nature. Now within this enduring fact we observe rest and motion. A body at rest in the space of our observation is tracing out a certain historical route intersecting the moments of our time-system in a sequence of instantaneous points. This route is what we mean by a point of the permanent space of our time-system. Thus each time-system has its own space with its own points, and these permanent points are loci of instantaneous points.

The paradoxes of relativity arise from the fact that we have not noticed that when we change our time-system we change the meaning of time, the meaning of space and the meaning of points of space (conceived as permanent).

Now the route of a small body at rest in the space of a time-system, that is to say, a point of that time-system, has a certain symmetry in respect to the successive instantaneous spaces of that system, which is expressed for us by the perception of lack of change of position. This symmetry is the basis of the definition of rectangularity.

If the body be at rest in the space of the time-system *t*, it is moving in a straight line in the space of another time-system *T*. This permanent straight line intersects any moment of *T*, say T_1 in an instantaneous straight line l_1 (say). Then l_1, is perpendicular to the series of instantaneous parallel planes in which the moments of system *t* intersect T_1. In other words the planes to which motion is perpendicular are the planes of intersection with the moments of that time-system for whose space that motion would be represented as rest.

We have thus defined both parallelism and perpendicularity without reference to congruence, but in terms of immediate data of perception. Furthermore, the parallelism of the moments of one time-system enables us to extend parallelism to time as also expressing the relation to each other of permanent points of the same time-system. It thus follows that we now possess a structure in terms of which congruence can be defined. This means that there will be a class of qualities *y* one and only one of which attaches to any stretch on a straight line or on a point, such that matching in respect to this quality is what we mean by congruence.

The thesis that I have been maintaining is that measurement presupposes a perception of matching in quality. Accordingly in examining the meaning of any particular kind of measurement we have to ask, What is the quality that matches?

Furthermore, in applying this doctrine to measurements in space and time, I have maintained that the things whose qualities match are events. In other words, I maintained that it is events that are congruent, and that spatial congruence and temporal congruence are merely special instances of this fundamental congruence. In conformity with this doctrine I also maintain that space and time are merely the exhibition of relations between events.

The usual opinion, or at any rate the more usual mode of expression, is that space and time are relations between the material objects implicated in events. It is difficult to understand how time can be a relation between two permanent objects. Also with the modern assimilation of time and space, this difficulty in respect to time also attaches to space. Furthermore, I hold that these permanent objects are nothing else than adjectives of events. It follows that a yard measure is merely a device for making evident the spatial congruence of the events in which it is implicated.

The divergence between the two points of view as to space-time, that is to say, as to whether it exhibits relations between events or relations between objects in events is really of the utmost importance in the stage of physical science. If it be a relatedness between events, it has the character of a systematic uniform relatedness between events which is independent of the contingent adjectives of events. In this case we must reject Einstein's view of a heterogeneity in space-time. But if space-time be a relatedness between objects, it shares in the contingency of objects, and may be expected to acquire a heterogeneity from the contingent character of objects. I cannot understand what meaning can be assigned to the distance of the sun from Sirius if the very nature of space depends upon casual intervening objects which we know nothing about. Unless we start with some knowledge of a systematically related structure of space-time we are dependent upon the contingent relations of bodies which we have not examined and cannot prejudge.

Furthermore, how time is to be got from the relations of permanent bodies completely puzzles me. And yet the moderns assimilate time with space. I have never seen even the beginning of an explanation of the meaning of the usual phraseology.

I have already reiterated, that measurement presupposes a structure yielding definite stretches which, in some sense inherent in the structure, match each other; and I have explained the type of structure which is formed by our space-time.

The essence of this structure is that it is stratified in many different ways by different time-systems. This is a very peculiar idea which is the product of the speculations of the last fifteen years or so. We owe the whole conception notably to Einstein. I do not agree with his way of handling his discovery. But I have no doubt as to its general correctness. It is at first sight somewhat of a shock to think that other beings may slice nature into time-sections in a different way to what we do. In fact we have differences even among ourselves which luckily are quite imperceptible. However if we allow this possibility we not only explain many modern delicate experiments, but we also obtain explanations of what we mean by the spatial extension in three dimensions, and by planes and straight lines, and parallels and right-angles. We also obtain a definite meaning for the matching which is the basis of our congruence. The explanation is too sweeping to be put aside. Our whole geometry is merely the expression of the ways in which different events are implicated in different time-systems.

I have also hitherto omitted to point out that all order in space is merely the expression of order in time. For a series of parallel planes in the space of our time is merely the series of intersections with a series of moments of another time-system. Thus the order of the parallel planes is merely the time-order of the moments of this other system.

I must stop now. We started from the simplest idea which meets every child at the beginning of his or her schooldays. I mean the idea of equality.

We asked what it meant. We have then been led on and on, till we have found ourselves plunged in the abstruse modern speculations concerning the character of the Universe. They are not really very difficult. I call them abstruse because they deal with questions which we do not ordinarily think about. It is therefore a strain on our imaginations to follow the line of thought. But when we have once allowed the possibility of different meanings for time in nature, the argument is a straightforward deduction of the consequences.

CHAPTER IV

SOME PRINCIPLES OF PHYSICAL SCIENCE

It is my ambition in this lecture to discuss some general principles of mathematical physics, and to illustrate them by their application to the problem of the gravitational field. In a sense such a discussion should form the first chapter of the science, but it is that first chapter which is studied last.

The Apparent World. It would be easy to quote an imposing array of authorities, almost a consensus of authorities, in support of the thesis that the subject matter of physical science is composed of things observed by the senses. Such things are sights, sounds, touches, bodily feelings, shapes, distances, and their mutual relations. I will call the whole assemblage of them the 'apparent world.' Natural science is therefore the study of the interconnections of the things forming the apparent world.

This profession of the motive of science seems however in sharp contradiction to its actual achievement. The molecular theory, the wave theory of light, and finally the electromagnetic theory of things in general have, as it seems, set up for scientific investigation a society of entities, such as ether, molecules, and electrons, which are intrinsically incapable of direct observation. When Sir Ernest Rutherford at Cambridge knocks a molecule to pieces, he does not see a molecule or an electron. What he observes is a flash of light. There is at most a parallelism between his observation and the conjectural molecular catastrophe.

I suggest to you that, unless we are careful in our formulation of principles, the outcome of this train of thought is apt to be unsatisfactory and very misleading to scientific imagination. The apparent world becomes an individual psychological reaction to the stimulus of an entirely disparate interplay of electrons and ether. The whole of it is in the same boat. There is no principle by which we can assign for some of it any independence of individual psychology superior to that of the remainder.

On this theory we must entirely separate psychological time, space, external perceptions, and bodily feelings from the scientific world of molecular interaction. This strange world of science dwells apart like the gods of Epicurus, except that it has the peculiar property of inducing our minds to play upon us the familiar antics of our senses.

If we are to avoid this unfortunate bifurcation, we must construe our knowledge of the apparent world as being an individual experience of something which is more than personal. Nature is thus a totality including individual experiences, so that we must reject the distinction between nature as it really is and experiences of it which are purely psychological. Our experiences of the apparent world are nature itself.

Two-fold Cognisance. We have a two-fold cognisance of nature, and I will name the two factors of this experience 'cognisance by adjective' and 'cognisance by relatedness.'

Think of yourself as saying, 'There is a red patch.' You are affirming redness of something, and you are primarily conscious of that something because of its redness. In other words, the redness exhibits to you the something which is red. This is cognisance by adjective, red being the adjective. But your experience has gone further than mere cognisance by adjective. Your knowledge is not merely of something which is red. The patch is *there* and it endures while you are observing it. Thus you are cognisant of it as having spatio-temporal position, and by this we mean a certain type of relatedness to the rest of nature which is thereby involved in our particular experience. This knowledge of nature arising from its interconnectedness by spatio-temporal relations is cognisance by relatedness.

For example, the physiological account of the function of the brain as determining the conditions of external perception presupposes that the events of the brain signify the totality of contemporaneous space. Again the disclosure of space behind the looking-glass as qualified by images situated in it exhibits the fact that the events in front of the glass are significant of contemporaneous space behind it. Also we know that there is space inside the closed cupboard.

Nature is an abstraction from something more concrete than itself which must also include imagination, thought, and emotion. This abstraction is characterised by the systematic coherency of its interconnections disclosed in cognisance by relatedness. Thus the substances of nature which have the cognised adjectives as their qualities are also the things in nature connected by the cognised relatedness. Nature is delimited as the field of this closed system of related things. Accordingly the ultimate facts of nature are events, and the essence of cognisance by relatedness is the ability to specify the event by time and by place. Dreams are ruled out by their inability to pass this test.

But an event can be specified in this way without its being the subject of direct cognisance by adjective. For example we can exactly specify a time and a place on the further surface of the moon, but we should very much like to know what is happening there. There is however a certain fullness in the dual cognisance both by adjective and by relatedness. I will use the term 'perception' for this full experience.

Mere cognisance by relatedness is essentially knowledge of an event merely by its spatio-temporal relations to other events which are perceived and thus form a framework of what is fully experienced. In this sense there is no cognisance by relatedness without perception.

It is not the case that the analysis of the adjectives of appearance attached to the events within any limited field of nature carries with it any certain knowledge of adjectives attached to other events in the rest of nature, or indeed of other such adjectives attached to those same events. I will refer to this fact by the phrase, the contingency of appearance.

On the other hand, though the character of time and space is not in any sense *a priori*, the essential relatedness of any perceived field of events to all other events requires that this relatedness of all events should conform to the ascertained disclosure derived from the limited field. For we can only know that distant events are spatio-temporally connected with the events immediately perceived by knowing what these relations are. In other words, these relations must possess a systematic uniformity in order that we may know of nature as extending beyond isolated cases subjected to the direct examination of individual perception. I will refer to this fact by the phrase, the uniform significance of events.

Thus the constitutive character of nature is expressed by 'the contingency of appearance' and 'the uniform significance of events.' These laws express characters of nature disclosed respectively in cognisance by adjective and cognisance by relatedness. This doctrine leads to the rejection of Einstein's interpretation of his formulae, as expressing a casual heterogeneity of spatio-temporal warping, dependent upon contingent adjectives.

The case of the yard-measure illustrates my meaning. It is a contingent adjective of the events where it is situated. Its spatio-temporal properties are entirely derived from the events which it qualifies. For example, its use depends on the recognition of simultaneity, so that we shall not observe its two ends at widely different times. But simultaneity concerns events. Also the mere self-identity of the yard-measure does not suffice for its use, since we also admit the continued identity of objects which shrink or expand. The yard-measure is merely a device for making evident obscure relations between those events in which it appears.

If congruence merely meant relations between contingent adjectives of appearance, there would be no measurement of spatial distance or of temporal lapse without knowledge of actual intervening appearances, and no meaning for such distance in the absence of these adjectives. For example, the 'distance of the star Sirius' would be a phrase without meaning.

You will have observed that in this doctrine of cognisance by relatedness I am merely taking the old belief that we know of unbounded time and of unbounded space and am adapting it to my inversion which gives the supremacy to events and reduces time and space to mere relations between them.

The Doctrine of Time. It follows from my refusal to bifurcate nature into individual experience and external cause that we must reject the distinction between psychological time which is personal and impersonal time as it is in nature. Two conclusions follow, of which the one is conservative, and indeed almost reactionary, and the other is paradoxical.

The conservative conclusion is that in cognisance by relatedness the apparent world is disclosed as stratified into a succession of strata which are subordinate totalities of immediate experience. Each short duration of time is merely a total slab of nature disclosed as a totality in cognisance by relatedness, and for any individual experience partially disclosed in cognisance by adjective. There can be no other meaning for time, if we admit the position from which my argument has started. I will state the doctrine in this way, Time is a stratification of nature. Adherence to this doctrine is today the mark of a reactionary. I accept the term with the qualification that it is reaction to the admission of obvious fact.

We now pass to the other conclusion which is paradoxical. The assumption of the uniqueness of the temporal stratification of nature has slipped into human thought. Certainly in each individual experience such uniqueness must be granted. But confessedly each individual experience is partial, and we cannot safely reason from partial experience to the limitation of the variety of nature. Accordingly the uniqueness of time succession for each of us does not guarantee its consistency for all.

At this point I put by urgent metaphysical questions concerning any supposed distinction between past, present, and future as to the character of their existence. Also I need not recall to your minds the reasons, based upon refined observations, for assuming the existence in nature of alternative time-systems entailing alternative systems of stratification.

I think that no one can study the evidence in its detail without becoming convinced that we are in the presence of one of the most profound reorganisations of scientific and philosophic thought. But so many considerations are raised, so diverse in character, that we are not justified in accepting blindfold the formulation of principles which guided Einstein to his formulae.

You will have observed that for reasons which I have briefly indicated, I maintain the old-fashioned belief in the fundamental character of simultaneity. But I adapt it to the novel outlook by the qualification that the meaning of simultaneity may be different in different individual experiences. Furthermore, since I start from the principle that what is apparent in individual experience is a fact of nature, it follows that there are in nature alternative systems of stratification involving different meanings for time and different meanings for space. Accordingly two events which may be simultaneous in one instantaneous space for one mode of stratification may not be simultaneous in an alternative mode.

Time and Space. The homogeneity of time with space arises from their common share in the more fundamental quality of extension which is a quality belonging exclusively to events. By extension I mean that quality in virtue of which one event may be part of another or two events may have a common part. Nature is a continuum of events so that any two events are both parts of some larger event.

The heterogeneity of time from space arises from the difference in the character of passage in time from that of passage in space. Passage is the same as significance, and by significance I mean that quality of an event which arises from its spatio-temporal relationships to other events.

For the sake of simplicity I will speak of events whose dimensions are ideally restricted. I will call them 'event-particles.' Also we may conceive of an event restricted except in one dimension. Such an event may be termed a route or path, where I am now thinking of a route of transition through the continuum of nature. A route may evidently be also conceived as a linear chain of event-particles. But its essential unity is thereby lost. A 'spatial' route is a route which lies entirely in one instantaneous space. A 'historical' route is such that no two of its event-particles are simultaneous according to any time-system. Along such a route there is a definite antecedence and subsequence in time which is independent of alternative time-systems.

Thus the distinction of time from space, which I have just asserted, consists in the fact that passage along a spatial route has a different character from passage along a historical route. For proof of this fact think of a spatial route which has a material particle situated in each of its event-particles. We pronounce at once that all these material particles are different, because no material particle can be in two places at the same time. But if a historical route is in like case and the material particles be of like character even with some differences, we equally pronounce them to be the same material particle at successive stages of its existence. This difference of judgment can only arise from the distinction in the characters of spatial and temporal passage.

It only strengthens this argument when we remember that the events are the ultimate substance of nature and that the apparent material particle is an adjective of appearance which qualifies them. For the unique type of individuality possessed by the emergence of the same adjective throughout the historical route must be due to the special peculiarity of the route. I will recur to this question later when I define adjectival particles.

Time-Systems. According to the view which I am urging on you a moment of time is to be identified with an instantaneous spread of the apparent world. The relations of interconnection within this moment form a momentary three-dimensional space. Such a space is an abstraction from the full-bloodedness of a moment of time which includes all that is apparent in that space. A time-system is a sequence of non-intersecting moments including all nature forwards and backwards. I call the moments of such a consistent system 'parallel' because all parallelism is derived from their mutual relations and from their intersections with the parallel moments of other time-systems.

I am also assuming on rather slight evidence that moments of different time-systems always intersect. This hypothesis is the simplest and I know of no phenomena that would be explained by its denial. The result is to introduce the peculiar properties of Euclidean parallelism.

One advantage of the admission of alternative time-systems is that they afford explanations of the notion of position and of the notion of evenly lying loci, such as planes and straight lines. However, I will not in this lecture enter into a detailed examination of the origins of geometry.

Permanent Space. The momentary spaces of a time-system are matters of direct observation, at least when we construe momentary in an approximate sense. They must be discriminated from the permanent space of that time-system.

Rest and motion are ultimate data of observation, and permanent space is the way of expressing the connections of these data. The ultimate elements of permanent space are therefore somewhat elaborate. It will be sufficient for my immediate purpose in this lecture to exhibit the meaning to be ascribed to a point of the permanent space of a time-system.

Consider observations wedded to a single temporal mode of stratification. Some apparent bodies will be observed to be in motion and others at rest. The historical route forming the successive situations of an apparent particle at rest for such an observer is a point in the permanent space which corresponds to that time-system. For an observer who is wedded to another time-system the same apparent particle will be moving with uniform velocity. Accordingly the point of the space of the former time-system cannot be a point of the space of the latter time-system, since to be at rest occupying one point in the space of one time-system is to be moving through a succession of points in the space of another time-system. A permanent point is thus highly complex and only serves for one particular specification of the meaning of space and time. Each event-particle will occur at one point in each permanent space, and is thus the vertex of a pencil of points, one point for each time-system.

Each point intersects any moment, of whatever time-system, in just one event-particle. There is thus a point-wise correlation between the event-particles of any momentary space and the points of the permanent space of any time-system. This correlation explains the naturalness with which observation of momentary spaces is expressed in terms of permanent space so as to gain the facile representation of the phenomena of rest and motion, which can have no existence for a single momentary space.

This general theory of the grounds in nature for geometry and time is consistent with a rigid relativity whereby space and time are simply expressions for a certain observed ordering of events. Also it is essential to note that the spatial relations between apparent bodies only arise mediately through their implication in events. It is essential to adopt this view if we are to admit any assimilation of space and time.

The Physical Field. We now pass to the consideration of the status of the physical field of natural science. The scope of the contingency of appearance is limited, and the conditions of limitation are what we term the laws of nature.

They are expressed by assuming that the apparent adjectives of the past indicate a certain distribution of character throughout events extending from the past into the future. It is further assumed that this hypothetical distribution of character in its turn expresses the possibilities of adjectives of appearance attachable to the future events. Thus the regulation of future adjectives of appearance by past adjectives of appearance is expressed by this intermediate distribution of character, indicated by the past and indicating the future.

I call this intermediate distribution of character the 'physical field.' The true expression of the physical field is always to some extent a matter of conjecture. The only guarantee for correctness is the pragmatic test that the theory works.

The physical field is not the cause of perception nor is it the object perceived. The search for a cause of perception raises a problem which is probably meaningless and certainly insoluble. The physical field is merely that character of nature which expresses the relatedness between the apparent adjectives of the past and the apparent adjectives of the future. It therefore shares in the contingency of appearance, and accordingly cannot affect spatio-temporal relations.

Atomicity. Luckily the physical field is atomic, so far as concerns our approximate measurements. By this I mean that we can discriminate in the four-dimensional continuum certain regions or events, such that each exhibits a physical character which is entirely independent of the physical characters of other events or of the other physical characters of that event. This physical character requires the whole region for its complete exposition. Thus atomicity implies two properties, one is the breakdown of relativity in that the atomic character is independent of the physical characters pervading the rest of nature, and the other is that we cannot completely exhibit this character without the whole corresponding region.

This physical atomic character is the only case in which the Aristotelian idea of an attribute of a substance holds without grave qualification, at least so far as the realm of nature is concerned. Furthermore, atomicity is a property which is capable of more or less complete realisation. Failure to attain complete atomicity is illustrated when one aspect of the physical field modifies another aspect of it, for example, when the physical field of mass modifies that of electro-magnetism.

Observe that the practical atomicity of the physical and apparent characters is essential for the intelligibility of the apparent world to a finite mind with only partial perception. Without atomicity we could not isolate our problems; every statement would require a detailed expression of all the facts of nature. It has always been a reproach to those philosophers who emphasize the systematic relatedness of reality that they make truth impossible for us by requiring a knowledge of all as a condition for a knowledge of any. In the account of nature which I have just given you this objection is met in two ways: In so far as nature is systematically related, it is a system of uniform relatedness; and in the second place, intelligibility is preserved amid the contingency of appearance by the breakdown of relatedness which is involved in atomicity.

This breakdown of relatedness in the expression of the laws of nature is reflected into observation by our perception of material objects. Such an object is more than its colour, is more than its touch, and is more than our feeling of its resistance to push. The object, taken throughout its history, is a permanent factor conditioning adjectives of appearance, and it is a factor which is largely independent of its relatedness to other contingent facts. It is the endeavour to make precise this aspect of a perceived material object which has led to the atomicity of modern science.

Thus it is not true to say, without qualification, that the physical field is not perceived. We do recognise permanences in the relatedness of things sensed, permanences which are largely disconnected. The physical field is the endeavour to express precisely these perceived permanences as atomic characters of events.

Adjectival Particles. The discussion of these recognised permanences is reduced to an ideal simplicity by the introduction of adjectival particles, by which I mean the ideally small perceived bodies and the elementary physical particles.

I have already stated that an adjectival particle receives its enduring individuality from the individuality of its historical route. Let me now give a more precise statement of my meaning: An 'adjectival particle' is the adjective attached to the separate event-particles of a historical route by virtue of the fact that some one and the same adjective attaches to *every* stretch of the route. It is the outcome of the transference to the individual event-particles of a common property of all the stretches.

Accordingly the unique individuality of the particle is nothing else than the fusion of the continued sameness of the adjective with the concrete individuality of the historical route. We must not think of an adjectival particle as moving through its route. We will say that it 'pervades' its route, and that it is 'situated' at each event-particle of the route, and that it 'moves' in an orbit in each permanent space.

It follows from this conception of the meaning of an adjectival particle that the expression of its properties should require the consideration of stretches of its route. In order, even now, to attain ideal simplicity we proceed to the limit of making all such stretches infinitesimally small. A stretch of a historical route, as thus employed in the process of proceeding to a limit, will be called a 'kinematic element.' A kinematic element is equivalent to both the position and the velocity of an adjectival particle in any permanent space at any time.

Mass-Particles. A mass-particle is an adjectival particle. It follows that for some limited purposes we can treat it as being situated in an event-particle, but that for the final purpose of enunciating the laws of nature we must conceive it as pervading a stretch of its historical route.

Consider [cf. figure, p. 31] first the former conception of a mass-particle m as situated at an event-particle which we will call P. The physical field due to m at P has to stretch away into the future. It is to be a limited atomic field with a foot in two camps, for it represents the property of the future as embodied in the past. It may therefore, so far as it is completely atomic, be expected to consist of that region within the future from P which has peculiar affinities with the region co-present with P.

Now what I call the kinematic future from P is the region traversed by the pencil of permanent points which has P as vertex, considering only the portions of those points which stream into the future from P. It will be remembered that there is one such point for each time-system. Again the region co-present with P is the region reached by the moments containing P. It will be remembered that each moment is an instantaneous three-dimensional space, and that there is one such moment for each time-system. Both these regions, the kinematic future from P and the region co-present with P, are four-dimensional. The ordered geometry of the four-dimensional continuum shows that the boundary region which separates the two is a three-dimensional region which belongs to neither. This three-dimensional region will be called the 'causal future' from P. It has all the properties that we want for an atomic region completely defined by P and for its delimitation not dependent upon any contingent characters of the rest of nature.

The atomic physical field of the mass-particle at P is P's causal future together with P itself. We will call P the origin of the field. The physical character of this field as a whole is what is meant by the mass-particle at P. This is merely Faraday's conception of the tubes of force as constituting the physical particle, with the modification that the tubes in the act of streaming through space also stream through time. Conceived under the guise of time and permanent space the mass-particle is a transmission of physical character along its lines of force with a definite finite velocity.

Metrical Formulae. A few mathematical formulae are now necessary for my argument. The assumption, adopted as the simplest representation of observed facts, that the permanent space of each time-system is Euclidean, leads to the formulae of the special theory of relativity. There is however this difference that the critical velocity c has no reference to light, and merely expresses the fact that a lapse of time and a stretch of spatial route can be congruent to each other.

Define the quantities

$$\omega_\mu, \left[\mu = 1, 2, 3, 4\right]$$

by

$$\left.\begin{aligned} \omega_\mu^2 &= 1, \left[\mu = 1, 2, 3,\right] \\ \omega_4^2 &= -c^2 \end{aligned}\right\} \quad \ldots\ldots\ldots\ldots\ldots (1).$$

Let a rectangular Cartesian system of coordinates in the permanent space of the 'x' time-system be (x_1, x_2, x_3) and let the lapse of x-time since zero time be x_4. Thus (x_1, x_2, x_3, x_4) are the four coordinates of an event-particle, which we will name X. Also in the 'y' time-system, we denote analogously a permanent point by the Cartesian coordinates (y_1, y_2, y_3) and a lapse of y-time by y_4. Let (y_1, y_2, y_3, y_4) and (x_1, x_2, x_3, x_4) denote the same event-particle.

Then [cf. *The Principles of Natural Knowledge,* Ch. XIII] the relations between the two systems of coordinates, the 'x' system and the 'y' system, are of the form

$$\omega_\mu\left(y_\mu - b_\mu\right) = \sum_\alpha l_{\mu\alpha}\,\omega_\alpha\, x_\alpha, \; \left[\mu = 1, 2, 3, 4\right] \cdots\cdots(2),$$

where the symbol $\sum_\alpha$ means summation for $\alpha = 1, 2, 3, 4$ successively, and the l's are constants satisfying the conditions

$$\left.\begin{aligned} \sum_\mu l_{\mu\alpha} l_{\mu\beta} &= 0, \; \left[\alpha \neq \beta\right] \\ &= 1, \; \left[\alpha = \beta\right] \end{aligned}\right\} \cdots\cdots\cdots\cdots\cdots(3),$$

These conditions entail analogous formulae for the converse transformation from 'y' to 'x.'

It follows that, if the coordinates of another event-particle, named P, be (p_1, p_2, p_3, p_4) in the 'x' system and (q_1, q_2, q_3, q_4) in the 'y' system,

$$-\sum_\alpha \omega_\alpha^{\,2}\left(x_\alpha - p_\alpha\right)^2 = -\sum_\mu \omega_\mu^{\,2}\left(y_\mu - q_\mu\right)^2.$$

Let $r_{(x)}$ and $r_{(y)}$ be respectively the x-distance and the y-distance between X and P. Then this invariant for X and P can be expressed indifferently either by

$$c^2\left(x_4 - p_4\right)^2 - r_{(x)}^{\,2}$$

or by

$$\left.c^2\left(y_4 - q_4\right)^2 - r_{(y)}^{\,2}\right\} \cdots\cdots\cdots\cdots\cdots(4).$$

Then

(i) X and P are co-present, if

$$c^2\,(x_4 - p_4)^2 - r_{(x)}^{\,2} < 0,$$

(ii) P is kinematically antecedent to X, if

$$x_4 > p_4, \text{ and } c^2(x_4 - p_4)^2 - r_{(x)}^{\,2} > 0,$$

(iii) X lies in the causal future from P, if

$$c(x_4 - p_4) = r_{(x)}.$$

Routes of Adjectival Particles. Let the mass-particle M be situated at X and the mass-particle m be situated at P, and let X' and P' be event-particles respectively neighbouring to X and P on the historical routes of M and m in the four-dimensional continuum of nature. Let their coordinates be respectively

$$(x_\mu + dx_\mu, \dots) \text{ and } (p_\mu + dp_\mu, \dots), \; [\mu = 1, 2, 3, 4].$$

These are accordingly infinitesimal invariants dG_M^2 and dG_m^2, respectively expressing a spatio-temporal property of the kinematic elements XX' and PP'. This property depends on the existence of the whole bundle of diverse time-systems without special emphasis on any one of them. These invariants [cf. equation (4)] are expressed by

$$\left.\begin{aligned} dG_M^2 &= -\sum_\alpha \omega_\alpha^2 dx_\alpha^2 \\ dG_m^2 &= -\sum_\alpha \omega_\alpha^2 dp_\alpha^2 \end{aligned}\right\} \dots\dots\dots\dots\dots (5).$$

Let the route of M be expressed by assuming x_1, x_2, x_3 to be appropriate functions of x_4, and the route of m by assuming p_1, p_2, p_3, to be appropriate functions of p_4. Thus, always in reference to these assumptions, we write

$$\dot{x}_\mu = \frac{dx_\mu}{dx_4}, \text{ and } \dot{p}_\mu = \frac{dp_\mu}{dp_4}.$$

Also we put

$$\left.\begin{aligned} v_m^2 &= \dot{p}_1^2 + \dot{p}_2^2 + \dot{p}_3^2 \\ v_M^2 &= \dot{x}_1^2 + \dot{x}_2^2 + \dot{x}_3^2 \end{aligned}\right\} \dots\dots\dots\dots\dots (6)$$

and

$$\left.\begin{aligned} \Omega_m &= \left\{1 - \frac{v_m^2}{c^2}\right\}^{-\frac{1}{2}} \\ \Omega_M &= \left\{1 - \frac{v_M^2}{c^2}\right\}^{-\frac{1}{2}} \end{aligned}\right\} \dots\dots\dots\dots\dots (7)$$

and

$$\xi_m = c^{-1}\left\{(x_1 - p_1)\dot{p}_1 + (x_2 - p_2)\dot{p}_2 + (x_3 - p_3)\dot{p}_3\right\}$$

Impetus. In order to exhibit the character of the physical field due to a mass-particle we must consider it as pervading a kinematic element, which has the advantage over an event-particle of retaining the quality of historic passage. A loss of spatial dimensions is comparatively immaterial, though it probably represents a simplification beyond anything which obtains in nature.

In expressing the physical field due to *m* we must therefore consider the kinematic element *PP′* of its route. Also we must take any arbitrary element *XX′*, and consider how its qualifications as a possible kinematic element of the route of *M* are affected by the fact that *m* pervades the element *PP′*.

Each kinematic element, such as *XX′*, having *X* as initial starting-point will have certain physical characters. The assemblage of quantities defining these physical characters for this pencil of elements constitutes the physical field at *X*. The two such characters which we need consider, as qualifying *XX′* for pervasion by *M*, are its potential mass impetus and its potential electromagnetic impetus.

The potential mass impetus along *XX′* will be written $\sqrt{dJ^2}$, and the potential electromagnetic impetus will be written *dF*. If the mass of the particle *M* be also denoted by *M*, and its electric charge, in electrostatic units, by *E*, then the realised mass impetus due to pervasion of *XX′* by *M* will be

$$M\sqrt{dJ^2},$$

and the realised electromagnetic impetus, due to the same pervasion, will be

$$c^{-1}\,EdF.$$

The total impetus along *XX′* realised by its pervasion by *M* is

$$dI + M\sqrt{dJ^2} + c^{-1}\,EdF \;\ldots\ldots\ldots\ldots\ldots\ldots\ldots\ldots\ldots\ldots (9).$$

Summing along the route of M between the assigned event-particles A to B, we obtain the realised impetus along this route which is symbolised by

$$\int_A^B \left\{ M\sqrt{dJ^2} + c^{-2}\,E\,dF \right\}.$$

If this total impetus is to be finite, it is evident that $\sqrt{dJ}$ and dF must be homogeneous functions of du_1, du_2, du_3, du_4 of the first degree, where (u_1, u_2, u_3, u_4) are any generalised coordinates of X. Thus, guided empirically by the ascertained character of dynamical equations and of the electromagnetic field, we can assume

$$\left.\begin{aligned} dJ^2 &= \sum_\mu \sum_\nu J_{\mu\nu}^{(u)}\, du_\mu du_\nu \\ dF &= \sum_\mu F_\mu^{(u)}\, du_\mu \end{aligned}\right\} \quad \ldots\ldots\ldots\ldots (10).$$

Thus $\| J_{\mu\nu}^{(u)} \|$ is a symmetric covariant tensor of the second order and $\| F_{\mu\nu}^{(u)} \|$ is a covariant tensor of the first order. The elements of these tensors are functions of the coordinates of X, that is, of (u_1, u_2, u_3, u_4). These tensors define the physical field at X so far as inertial and electromagnetic properties are concerned.

Hence, writing as above

$$\dot{u}_\mu = \frac{du_\mu}{du_4}, \quad [\mu = 1, 2, 3, 4]$$

for differentiation along the route of M, $\frac{dI}{du_4}$ is a function of $\dot{u}_1$, $\dot{u}_2$, $\dot{u}_3$, and of u_1, u_2, u_3, u_4. We now assume that the actual route of M satisfies the condition that the realised impetus is stationary between A and B for small variations of route. We thus obtain the equations of motion

$$\frac{d}{du_4}\frac{\partial}{\partial \dot{u}_\mu}\frac{dI}{du_4} - \frac{\partial}{\partial u_\mu}\frac{dI}{du_4} = 0, \quad [\mu = 1, 2, 3] \quad \ldots\ldots\ldots (11).$$

Expression for the Gravitational Field. I will now confine myself to the proper determination of dJ^2, as affected by the existence of other mass-particles m, m', etc., in other routes. In expressing the conditions restraining the contingency of appearance it is necessary that we have recourse to that aspect of nature which is independent of this contingency. The only such aspect is that arising from spatio-temporal properties. Also dG_M^2 and dG_m^2 are the invariants expressing the quantitative aspect of the historical passage of the elements XX' and PP'.

Again in considering the physical character of XX' as affected by m in its route, we must select that kinematic element PP' of m's route which is causally correlated with XX'. By this I mean that PP' has a point-wise correlation with XX' such that X is in the causal future from P and X' is in the causal future from P'. With this correlation the physical character of PP' is already determined when XX' occurs.

This assumption of causal correlation is mathematically expressed by the relation

$$x_4 - p_4 = r_{(x)}/c \quad \text{..................................} (12)$$

between corresponding event-particles on XX' and PP'.

The main empirical facts of gravitation are expressed by the assumption that

$$dJ^2 = dG_M^2 - \frac{2}{c^2}\sum_m \Psi_m \, dG_m^2$$

where $\sum_m$ means the summation for all mass-particles such as m in kinematic elements such as PP', causally correlated to XX', and Ψ_m expresses the gravitational law of fading intensity. The factor $2/c^2$ is inserted so that, when the main intensity is empirically adjusted to give the main inverse square law of gravitation, Ψ_m may be the analogue of the familiar gravitational potential at X due to m. It is easy to prove [cf. Part III] that, apart from any assumption of causal correlation between X and P,

$$\Omega_m\left\{c\left(x_4 - p_4\right) - \xi_m\right\}$$

has an invariant value for all sets of rectangular Cartesian coordinates in all time-systems. Also with the causal correlation between PP' and XX' which we are assuming, this invariant expression reduces to

$$\Omega_m\left\{r_{(x)} - \xi_m\right\}.$$

Accordingly, guided by our knowledge of the Newtonian law of gravitation, we assume

$$\Psi_m = \frac{\gamma m}{\Omega_m\left\{r_{(x)} - \xi_m\right\}} \quad \ldots\ldots\ldots\ldots(14),$$

where γ is the familiar constant of gravitation so as to produce the scale of intensity of the main inverse square Newtonian term.

If we write

$$\Psi = \sum_m \frac{\gamma m}{\Omega_m\left\{r_{(x)} - \xi_m\right\}} \quad \ldots\ldots\ldots\ldots(15),$$

then in an empty region Ψ satisfies

$$\nabla^2_{(x)}\Psi - \frac{1}{c^2}\frac{\partial^2\Psi}{\partial x_4^{\,2}} = 0 \quad \ldots\ldots\ldots\ldots(16).$$

We might, if we had preferred to do so, have started from the differential equation as the only invariant form of linear differential equation of the second order, and then deduced the above solution for Ψ_m as the only invariant solution for a single point-wise discontinuity. The procedure of thought which I have adopted seems to me to be better suited to throw into relief the fundamental ideas concerning nature.

Comparison with Einstein's Law. In the formula

$$dJ^2 = dG_M^{\,2} - \frac{2}{c^2}\sum_m \frac{\gamma m}{\Omega_m\left(r_{(x)} - \xi_m\right)} dG_m^{\,2} \quad \ldots\ldots\ldots\ldots(17)$$

$\sqrt{dJ^2}$ corresponds to Einstein's proper time*ds*. By identifying the potential mass impetus of a kinematic element with a spatio-temporal measurement Einstein, in my opinion, leaves the whole antecedent theory of measurement in confusion, when it is confronted with the actual conditions of our perceptual knowledge. The potential impetus shares in the contingency of appearance. It therefore follows that measurement on his theory lacks systematic uniformity and requires a knowledge of the actual contingent physical field before it is possible. For example, we could not say how far the image of a luminous object lies behind a looking-glass without knowing what is actually behind that looking-glass.

The above formula, assumed for dJ^2, also differs from Einstein's. In his procedure the J's are conditioned by making them satisfy the contracted Reimann-Christoffel tensor equations. He obtains a solution of these equations for a single point-singularity under the assumption that the gravitational field is permanent for the coordinates adopted so that no elements of the array $\| J_{\mu\nu} \|$ are functions of the time in the system of coordinates adopted. This limitation rules out any application of this solution to cases like that of the moon's motion, where the sun and earth evidently cannot both produce gravitational fields permanent for the same system of coordinates. My formula, given above, applies generally to all such cases. It is a matter for investigation whether the small terms depending on the motions thereby introduced into the gravitational formulae produce effects which are verified in observation as recorded in the discrepancies of the moon's tables. I have traced some theoretical effects of these terms of the order of magnitude of one or two seconds of arc with periods of the order of a month or a year, but I have not yet succeeded in hitting on a term of a period long enough to aggregate an observable effect, having regard to the state of the moon's tables. We want periods of about 250 years.

If the above formula gives results which are discrepant with observation, it would be quite possible with my general theory of nature to adopt Einstein's formula, based upon his differential equations, for the determination of the gravitational field. They have however, as initial assumptions, the disadvantage of being difficult to solve and not linear. But it is purely a matter for experiment to decide which formula gives the small corrections which are observed in nature. So far as matters stand at present both formulae give the motion of Mercury's perihelion, my formula gives a possible shift of the spectral lines dependent upon the structure of the molecule and on the interplay of the gravitational and electromagnetic fields, and lastly, assuming a well-known modification of Maxwell's equations giving such an interplay, the famous eclipse results follow[1].

Alternative Laws of Gravitation. Perhaps neither of the above formulae will survive further tests of other delicate observations. In this event we are not at the end of our resources. There are, in addition to Einstein's, yet two other sets of tensor differential equations which on the theory of nature explained in this lecture satisfy all the general requirements. These requirements are (i) to have no arbitrary reference to any one particular time-system, and (ii) to give the Newtonian term of the inverse square law, and (iii) to yield the small corrections which explain various residual results which cannot be deduced as effects of the main Newtonian law.

The possibility of other such laws, expressed in sets of differential equations other than Einstein's, arises from the fact that on my theory there is a relevant fact of nature which is absent on Einstein's theory. This fact is the whole bundle of alternative time stratifications arising from the uniform significance of events. It is expressed, without emphasis on any one such time-system, by the Galilean tensor $\|G_{\mu\nu}^{(u)}\|$. This tensor is defined by the property that, when expressed in terms of rectangular Cartesian coordinates (x_1, x_2, x_3, x_4) for any time-system 'x,'

$$\left.\begin{array}{l} G_{\mu\nu}^{(x)} = 0, \ [\mu \neq \nu] \\ G_{\mu\mu}^{(x)} = -\omega_u^{\ 2}, \ [\mu = 1, 2, 3, 4] \end{array}\right\} \text{.......................} (18).$$

Thus we have on hand two tensors, the above Galilean tensor and the tensor of the gravitational field which is

$$\|J_{\mu\nu}^{(x)}\|.$$

In order to formulate the differential equations involving the gravitational laws we shall require the three-index symbols of the first and second types for both the tensors $\|J_{\mu\nu}\|$ and $\|G_{\mu\nu}\|$. They will be written[2]

$$J\left[\mu\nu, \lambda\right]^{(u)} \text{ and } G\left[\mu\nu, \lambda\right]^{(u)}$$

for the symbols of the first type, and

$$J\left\{\mu\nu,\ \lambda\right\}^{(u)} \text{ and } G\left\{\mu\nu,\ \lambda\right\}^{(u)}$$

for the symbols of the second type. Also the associate contravariant tensors are written $\| J^{\mu\nu}_{(u)} \|$ and $\| G^{\mu\nu}_{(u)} \|$, and the determinant $\| J^{(u)}_{\mu\nu} \|$ is symbolised by $J^{(u)}$.

(i) Einstein's Law is

$$\left.\begin{aligned}
&\sum_{\rho}\left[\frac{\partial}{\partial u_{\rho}} J\left\{\mu\nu,\ \rho\right\}^{(u)} + J\left\{\mu\nu, \rho\right\}^{(u)} \frac{\partial}{\partial u_{\rho}} \log\left\{-J^{(u)}\right\}^{\frac{1}{2}}\right] \\
&- \sum_{\rho}\sum_{\sigma} J\left\{\mu\sigma,\ \rho\right\}^{(u)} J\left\{\nu\rho,\ \sigma\right\}^{(u)} - \frac{\partial^{2}}{\partial u_{\mu}\partial u_{\nu}} \log\left\{-J^{(u)}\right\}^{\frac{1}{2}} = 0, \\
&\qquad\qquad\qquad\qquad\qquad\qquad \left[\mu, \nu = 1,\ 2,\ 3,\ 4\right]
\end{aligned}\right\}$$

$$\ldots\ldots\ldots(19).$$

The two other laws which involve differential equations depend upon making the proper substitutions for the mixed tensor

$$\| K^{\lambda}_{\mu\nu} \|$$

in the following tensor equations

$$\left.\begin{aligned}
&\sum_{\rho}\left[\frac{\partial K^{\rho}_{\mu\nu}}{\partial u_{\rho}} + K^{\rho}_{\mu\nu} \frac{\partial}{\partial u_{\rho}} \log\left\{-G^{(u)}\right\}^{\frac{1}{2}}\right] \\
&\quad - \sum_{\rho}\sum_{\sigma}\left[K^{\sigma}_{\rho\nu} G\left\{\mu\sigma,\ \rho\right\}^{(u)} + K^{\sigma}_{\rho\mu} G\left\{\nu\sigma,\ \rho\right\}^{(u)}\right] = 0, \\
&\qquad\qquad\qquad\qquad\qquad\qquad \left[\mu, \nu = 1,\ 2,\ 3,\ 4\right]
\end{aligned}\right\} \ldots\ldots(20).$$

(ii) In this law the mixed tensor $\| K^{\lambda}_{\mu\nu} \|$ of the above equation is to stand for

$$\left\| \sum_{\alpha} G^{\lambda\alpha}_{(u)} J\left[\mu\nu, \alpha\right]^{(u)} - \sum_{\alpha}\sum_{\beta} G^{\lambda\alpha}_{(u)} J^{(u)}_{\alpha\beta} G\left\{\mu\nu, \beta\right\}^{(u)} \right\|.$$

(iii) In this law the mixed tensor $\| K^{\lambda}_{\mu\nu} \|$ of the equation above is to stand for,

$$\left\| \sum_{\alpha} T^{\lambda\alpha}_{(u)} J\left[\mu\nu, \alpha\right]^{(u)} - \sum_{\alpha}\sum_{\beta} T^{\lambda\alpha}_{(u)} J^{(u)}_{\alpha\beta} G\left\{\mu\nu, \beta\right\}^{(u)} \right\|,$$

where $\| T^{\mu\nu}_{(u)} \|$ is some contravariant tensor arising from some quality of the electromagnetic field. This law is suited to express the interaction (if any) of the electromagnetic field on the gravitational field.

If the equations of laws (ii) and (iii) be referred to rectangular Cartesian coordinates, they become

$$\text{(ii)} \qquad \sum_{\rho} \frac{1}{\omega_{\rho}^{2}} \frac{\partial}{\partial x_{\rho}} J\left[\mu\nu, \rho\right]^{(x)} = 0, \left[\mu, \nu = 1, 2, 3, 4\right] \quad ..(21),$$

and

$$\text{(iii)} \qquad \sum_{\rho}\sum_{\sigma} \frac{\partial}{\partial x_{\rho}} T^{\rho\sigma}_{(x)} J\left[\mu\nu, \sigma\right]^{(x)} = 0, \quad \left[\mu, \nu = 1, 2, 3, 4\right] \; (22).$$

(iv) The fourth law has already been considered. It can be expressed in the integral form

$$dJ^{2} = dG_{M}^{2} - \frac{2}{c^{2}} \sum_{m} \frac{\gamma m}{\Omega_{m}\left(r_{(x)} - \xi_{m}\right)} dG_{m}^{2} \; \ldots\ldots\ldots..(23),$$

where the kinematic element corresponding to dG_m^2 is causally correlated to that corresponding to dG_M^2.

According to this law the fundamental character of inertial properties is derived from their intimate connection with the abstract measure of uniform process in the spatio-temporal field. Thus $\sqrt{dG_M^2}$ and $\sqrt{dG_m^2}$ are these abstract measures of spatio-temporal process in the elements XX' and PP' of the tracks of M and m respectively. The inertial physical field modifies this abstract measure of process into the more concrete potential impetus $\sqrt{dJ^2}$, and full concreteness, so far as it is ascribable to nature, is obtained in the realised impetus $M\sqrt{dJ^2}$.

Rotation. In conclusion I will for one moment draw your attention to rotation. The effects of rotation are among the most widespread phenomena of the apparent world, exemplified in the most gigantic nebulae and in the minutest molecules. The most obvious fact about rotational effects are their apparent disconnections from outlying phenomena. Rotation is the stronghold of those who believe that in some sense there is an absolute space to provide a framework of dynamical axes. Newton cited it in support of this doctrine. The Einstein theory in explaining gravitation has made rotation an entire mystery. Is the earth's relation to the stars the reason why it bulges at the equator? Are we to understand that if there were a larger proportion of run-away stars, the earth's polar and equatorial axes would be equal, and that the nebulae would lose their spiral form, and that the influence of the earth's rotation on meteorology would cease? Is it the influence of the stars which prevents the earth from falling into the sun? The theory of space and time given in this lecture, with its fundamental insistence on the bundle of time-systems with their permanent spaces, provides the necessary dynamical axes and thus accounts for these fundamental phenomena. I hold this fact to be a strong argument in its favour, based entirely on the direct results of experience.

Conclusion. The course of my argument has led me generally to couple my allusions to Einstein with some criticism. But that does not in any way represent my attitude towards him. My whole course of thought presupposes the magnificent stroke of genius by which Einstein and Minkowski assimilated time and space. It also presupposes the general method of seeking tensor or invariant relations as general expressions for the laws of the physical field, a method due to Einstein. But the worst homage we can pay to genius is to accept uncritically formulations of truths which we owe to it.

PART II

PHYSICAL APPLICATIONS

CHAPTER V

THE EQUATIONS OF MOTION

THE equations of motion of a mass-particle *(M)* are [cf. Chapter IV, equation (11)]

$$\frac{d}{du_4}\frac{\partial}{\partial \dot{u}_\mu}\frac{dI}{du_4} - \frac{\partial}{\partial u_\mu}\frac{dI}{du_4} = 0, \left[\mu = 1, 2, 3\right] \quad \ldots\ldots(1),$$

where [cf. Chapter IV, equation (9)]

$$dI = M\sqrt{dJ^2} + c^{-1}\,E\,dF \quad \ldots\ldots(2).$$

We write

$$dJ^2 = dG_M^{\ 2} - \sum_\mu \sum_\nu g_{\mu\nu}^{(u)}\,du_\mu\,du_\nu \quad \ldots\ldots(3),$$

$$\Gamma_{(u)} = c^{-1}\sqrt{\frac{dJ^2}{du_4^{\ 2}}} \quad \ldots\ldots(4),$$

$$F_{\mu\rho}^{(u)} = \frac{\partial F_\mu^{\ (u)}}{\partial u_\rho} - \frac{\partial F_\rho^{\ (u)}}{\partial u_\mu} \quad \ldots\ldots(5).$$

Then the equations of motion can be written

$$\frac{d}{du_4}\left\{\frac{M}{\Gamma_{(u)}}\frac{\partial}{\partial \dot{u}_\mu}\left(-\tfrac{1}{2}c^2\Gamma_{(u)}^{\ 2}\right)\right\} - \frac{M}{\Gamma_{(u)}}\frac{\partial}{\partial u_\mu}\left(-\tfrac{1}{2}c^2\Gamma_{(u)}^{\ 2}\right)$$

$$= E\sum_\rho F_{\mu\rho}^{(u)}\,\dot{u}_\rho, \quad \left[\mu = 1, 2, 3\right] \ldots(6),$$

where (u_1, u_2, u_3, u_4) are any generalised coordinates of the situation of M.

If (x_1, x_2, x_3, x_4) are Cartesian coordinates for the spatio-temporal system 'x,' these equations become

$$\frac{d}{dx_4}\frac{M\dot{x}_\mu}{\Gamma} + \sum_\rho g_{\mu\rho}\frac{d}{dx_4}\frac{M\dot{x}_\rho}{\Gamma} + \frac{M}{\Gamma}\sum_\rho\sum_\sigma g\left[\rho\sigma,\ \mu\right]^{(x)}\dot{x}_\rho\,\dot{x}_\sigma$$
$$= E\sum_\rho F^{(x)}_{\mu\rho}\,\dot{x}_\rho,\ \left[\mu = 1, 2, 3\right] \quad \text{................(7).}$$

where

$$g\left[\rho\sigma,\ \mu\right]^{(x)} = \frac{1}{2}\left(\frac{\partial g^{(x)}_{\mu\rho}}{\partial x_\sigma} + \frac{\partial g^{(x)}_{\mu\sigma}}{\partial x_\rho} - \frac{\partial g^{(x)}_{\rho\sigma}}{\partial x_\mu}\right) \quad \text{............(8),}$$

and Γ is written for $\Gamma_{(x)}$. We write $\sum'_\rho$ for summation for $\rho = 1, 2, 3$, excluding $\rho = 4$. Then the terms

$$\frac{M}{\Gamma}\sum'_\rho\sum'_\sigma g\left[\rho\sigma,\ \mu\right]^{(x)}\dot{x}_\rho\,\dot{x}_\sigma$$

are called the 'pure centrifugal gravitational' terms, the terms

$$\frac{2M}{\Gamma}\sum'_\rho g\left[\rho 4,\ \mu\right]^{(x)}\dot{x}_\rho$$

are called the 'composite centrifugal gravitational' terms, and the term

$$\frac{M}{\Gamma}g\left[44,\ \mu\right]^{(x)}$$

is the 'pure gravitational' term. Also

$$\left(F^{(x)}_{14},\ F^{(x)}_{24},\ F^{(x)}_{34}\right)$$

is the electric force [electrostatic units] and

$$\left(cF^{(x)}_{23},\ cF^{(x)}_{31},\ cF^{(x)}_{12}\right)$$

is the magnetic force.

It is convenient to note for future reference that [cf. Chapter IV, equations (5) and (7)]

$$\left.\begin{aligned} dG_M^{\ 2} &= c^2\Omega_M^{\ -2}dx_4^{\ 2} \\ dG_m^{\ 2} &= c^2\Omega_m^{\ -2}dp_4^{\ 2} \end{aligned}\right\} \quad \ldots\ldots\ldots\ldots\ldots (9).$$

Also if

$$c\left(x_4 - p_4\right) = r_{(x)},$$

then

$$dp_4 = dx_4 - \frac{dr_{(x)}}{c} \quad \ldots\ldots\ldots\ldots\ldots (10).$$

CHAPTER VI

ON THE FORMULA FOR dJ^2

We adopt the formula [cf. Chapter IV, equation (17)]

$$dJ^2 = dG_M^2 - \frac{2}{c^2}\sum_m \frac{\gamma m}{\Omega_m (r-\xi)} dG_m^2 \quad \text{.................(1)},$$

where

$$c\left(x_4 - p_4\right) = r \quad \text{.................................(2)},$$

and m *is a* typical member of the attracting particles, situated at (p_1, p_2, p_3, p_4), and r stands for $r_{(x)}$ in Chapter IV.

Then [cf. equations (3), (9) and (10) of Chapter v]

$$\left.\begin{aligned}\sum_\mu \sum_\nu g^{(x)}_{\mu\nu} dx_\mu dx_\nu &= \frac{2}{c^2}\sum_m \frac{\gamma m}{\Omega_m (r-\xi)} dG_m^2 \\ &= \sum_m \frac{2\gamma m}{\Omega_m^3 (r-\xi)}\left(dx_4 - \frac{dr}{c}\right)^2\end{aligned}\right\} \quad \text{.........(3)}.$$

Thus

$$\left.\begin{aligned} g^{(x)}_{\mu\nu} &= \sum_m \frac{2\gamma m}{c^2 \Omega_m^3 (r-\xi)} \frac{\partial r}{\partial x_\mu}\frac{\partial r}{\partial x_\nu}, \quad [\mu, \nu \neq 4] \\ g^{(x)}_{\mu 4} &= \sum_m \frac{2\gamma m}{c^2 \Omega_m^3 (r-\xi)} \frac{\partial r}{\partial x_\mu}\left(\frac{\partial r}{\partial x_4} - c\right), \quad [\mu, \neq 4] \\ g^{(x)}_{44} &= \sum_m \frac{2\gamma m}{c^2 \Omega_m^3 (r-\xi)}\left(\frac{\partial r}{\partial x_4} - c\right)^2 \end{aligned}\right\} \quad \text{.......(4)}.$$

Also

$$\left.\begin{aligned}
&\frac{\partial r}{\partial x_\mu} = \frac{x_\mu - p_\mu}{r - \xi}, \quad [\mu \neq 4] \\
&\frac{\partial p_4}{\partial x_4} = 1 - \frac{1}{c}\frac{\partial r}{\partial x_4} = \frac{r}{r-\xi}, \quad \frac{\partial r}{\partial x_4} = \frac{-c\xi}{r-\xi} \\
&\frac{\partial p_4}{\partial x_\mu} = -\frac{1}{c}\frac{\partial r}{\partial x_\mu}
\end{aligned}\right\} \quad \ldots\ldots\ldots(5).$$

Also

$$\left.\begin{aligned}
&\frac{\partial \xi}{\partial x_4} = \frac{1}{c}\left\{1 - \frac{1}{c}\frac{\partial r}{\partial x_4}\right\}\left[\sum_\mu{}' \left(x_\mu - p_\mu\right)\ddot{p}_\mu - v_m^2\right] \\
&\frac{\partial \xi}{\partial x_\mu} = \frac{\dot{p}_\mu}{c} - \frac{1}{c^2}\frac{\partial r}{\partial x_\mu}\left[\sum_\mu{}' \left(x_\mu - p_\mu\right)\ddot{p}_\mu - v_m^2\right]
\end{aligned}\right\} \ldots\ldots\ldots(6).$$

The Potentials. It is convenient to express the components of $\| J_{\mu\nu} \|$ in terms of various potential functions which have either a tensor or an invariant character for transformations between space-time systems. We will limit our statements to rectangular Cartesian coordinates.

(i) *The General Potential.* This is symbolised by Φ, where

$$\Phi = \sum_m \frac{\gamma m}{\Omega_m (r - \xi)}\left[1 + \frac{1}{c^2}\Omega_m^2 \sum_\mu{}' \left(x_\mu - p_\mu\right)\ddot{p}_\mu\right]$$

$$- \frac{1}{c^3}\sum_m \gamma m \Omega_m^3 \sum_\mu{}' \dot{p}_\mu \ddot{p}_\mu \ldots.(7)$$

Here, as elsewhere, it is to be noticed that (x_1, x_2, x_3, x_4) lies in the causal future of (p_1, p_2, p_3, p_4) where m is situated; so that

$$c\left(x_4 - p_4\right) = r.$$

This condition always holds unless a special exception is made.

Φ is invariant. For

$$\Omega_m \left(r - \xi\right)$$

is invariant. Also, dropping for the moment the causal relation between

$$\left(x_1, x_2, x_3, x_4\right) \text{ and } \left(p_1, p_2, p_3, p_4\right),$$

$$\Omega_m\left\{c\left(x_4 - p_4\right) - \xi\right\} \text{ and } \Omega_m \frac{d}{dp_4}$$

are invariant. Hence

$$\Omega_m \frac{d}{dp_4}\left[\Omega_m\left\{c\left(x_4 - p_4\right) - \xi\right\}\right]$$

is invariant. Hence, replacing the causal relation after differentiation, we immediately find that

$$\Omega_m{}^2 \Sigma'_\mu \left(x_\mu - p_\mu\right) \ddot{p}_\mu - \frac{1}{c} \Omega_m{}^4 \left(r - \xi\right) \Sigma'_\mu \dot{p}_\mu \ddot{p}_\mu$$

is invariant. The invariance of Φ immediately follows.

(ii) *The Tensor Potential.* This is a covariant tensor of the first order, symbolised by

$$\left\| \Psi_\mu \right\|,$$

where

$$\Psi_\mu = -\Sigma_m \frac{\gamma m \omega_\mu{}^2 \dot{p}_\mu}{c^2\left(r - \xi\right)}$$

The tensor property follows from the fact that

$$\Omega_m\left(r - \xi\right) \quad \text{...(8)}$$

is invariant, and that

$$\left\| \Omega_m \omega_\mu{}^2 \dot{p}_\mu \right\|$$

is a covariant tensor.

We note that $$\Psi_4 = \Sigma_m \frac{\gamma m}{r - \xi} \quad \text{......................................(9).}$$

(iii) *The First Associate Potential.* This is symbolised by A, where

$$A = \Sigma_m \frac{\gamma m}{c^2} \Omega_m\left(r - \xi\right) \quad \text{..........................(10).}$$

It is obvious that A is invariant.

(iv) *The Second Associate Potential.* This is symbolised by B, where

$$B = \sum_m \frac{\gamma m}{c} \log \left\{ \Omega_m \left(r - \xi \right) \right\} \quad \text{.........................(11)}$$

It is obvious that B is invariant.

Then, neglecting terms involving c^3 as a factor,

$$dJ^2 = \left(c^2 + 6\Phi \right) dx_4{}^2 - \left(1 + \frac{2}{c^2} \Psi_4 \right) \sum_\mu{}' \, dx_\mu{}^2 - 8 \sum_\rho \Psi_\rho \, dx_\rho dx_4$$

$$+ 4 \sum_\rho \frac{\partial B}{\partial x_\rho} dx_\rho dx_4 + 2 \sum_\rho \sum_\sigma \frac{\partial^2 A}{\partial x_\rho \partial x_\sigma} dx_\rho dx_\sigma \quad \text{.......(12).}$$

It is now easy to transform to any other pure spatial coordinates in the 'x'-space by noting that

$$\left(\Psi_1^{(x)}, \Psi_2^{(x)}, \Psi_3^{(x)} \right)$$

is a vector in the *x-space,* and that

$$\sum_\rho \frac{\partial B}{\partial x_\rho} dx_\rho = dB,$$

and that, in any coordinate system (u_1, u_2, u_3, u_4),

$$\left\| \frac{\partial^2 A}{\partial u_\mu \partial u_\nu} - \sum_\rho \frac{\partial A}{\partial u_\rho} G\left\{ \mu\nu, \rho \right\}^{(u)} \right\|$$

is a covariant tensor of the second order, whatever may be the coordinates (u_1, u_2, u_3, u_4). Here[1] $G\{\mu\nu, \rho\}^{(u)}$ is the Christoffel three-index symbol defined by

$$G\left\{ \mu\nu, \rho \right\}^{(u)} = \sum_\alpha G_{(u)}^{\rho\alpha} \, G\left[\mu\nu, \alpha \right]^{(u)} \quad \text{..................(13).}$$

Furthermore, to our order of approximation [i.e. neglecting terms involving c^{-3}], the terms involving B will disappear from the equations of motion. Accordingly from this source no terms arise in these equations which involve c^{-1}. We shall also show that no terms of this order of magnitude arise from Φ, since these terms disappear from the approximate expression for Φ.

The Contemporary Positions. It is often more convenient to express the formulae in terms of the positions of the attracting particles in the x-space contemporary with the event-particle (x_1, x_2, x_3, x_4). Let this contemporary position of the particle m be (q_1, q_2, q_3, x_4).

We have to assume that r/c is a small time. Then

$$f(p_4) = f\left(x_4 - \frac{r}{c}\right)$$

$$= f(x_4) - \frac{r}{c} f'(x_4) + \frac{r^2}{2c^2} f''(x_4) \quad \text{..........(14)}.$$

Let R be the x-distance between (x_1, x_2, x_3) and (q_1, q_2, q_3), and

$$\left.\begin{aligned} \xi_0 &= \frac{1}{c}\sum_\mu{}' (x_\mu - q_\mu)\dot{q}_\mu \\ u_m^2 &= \sum_\mu{}' \dot{q}_\mu^2 \end{aligned}\right\} \quad \text{..........(15)}.$$

Then [cf. equation (14)]

$$\left.\begin{aligned} r - \xi &= R + \frac{1}{2}\frac{\xi_0^2}{R} - \frac{R}{2c^2}\left[u_m^2 - \sum_\mu{}'(x_\mu - q_\mu)\ddot{q}_\mu\right] \\ \xi &= \xi_0 + \frac{R}{c^2}\left\{u_m^2 - \sum_\mu{}'(x_\mu - q_\mu)\ddot{q}_\mu\right\} \end{aligned}\right\} \quad \text{........(16)}.$$

We write

$$\Omega^\circ, \ \Psi_\mu^\circ, \ A^\circ, \ B^\circ$$

for sufficient approximations to the various potential functions, neglecting terms involving c^{-3}.

Then

$$\Omega^\circ = \sum_m \frac{\gamma m}{R}\left[1 - \frac{1}{2}\frac{\xi_0^2}{R^2} + \frac{1}{2c^2}\sum_\mu{}'(x_\mu - q_\mu)\ddot{q}_\mu\right] \quad \text{..........(17)},$$

$$\left.\begin{aligned}\Psi_\mu^{\,o} &= -\sum_m \frac{\gamma m\, q_\mu}{c^2 R}, \quad [\mu = 1, 2, 3] \\ \Psi_4^{\,o} &= -\sum_m \frac{\gamma m}{R}\left[1 - \frac{1}{2}\frac{\xi_o^{\,2}}{R^2} + \frac{u_m^{\,2}}{2c^2} - \frac{1}{2c^2}\sum_\mu{}' \left(x_\mu - q_\mu\right)\ddot{q}_\mu\right]\end{aligned}\right\} \text{..............(18)},$$

$$A^o = \sum_m \frac{\gamma m}{c^2} R,$$

$$B^o = \sum_m \frac{\gamma m}{c} \log R \text{ ....(19)}.$$

We note that $$\nabla_{(x)}^{\;2} A^o = \frac{2}{c^2}\sum_m \frac{\gamma m}{R} \text{ ....(20)},$$

where $$\nabla_{(x)}^{\;2} = \frac{\partial^2}{\partial x_1^{\,2}} + \frac{\partial^2}{\partial x_2^{\,2}} + \frac{\partial^2}{\partial x_3^{\,2}}.$$

It easily follows [cf. equation (20)] that, if the attracting matter be a uniform sphere in the x-space of mass M' and of radius α, then at a distance R_o from the centre of the sphere and for points outside the sphere [i.e. $R_o > \alpha$]

$$\sum_m \frac{\gamma m}{R} = \gamma M'/R_o \text{ ....(21)},$$

$$A^o = \frac{\gamma M'}{c^2 R_o}\left\{R_o^{\,2} + \tfrac{1}{5}a^2\right\} \text{ ....(22)},$$

$$B^o = \frac{\gamma M'}{c}\log R_o + \frac{3\gamma M'}{2c}\sum_{s=1}^{\infty} \frac{1}{s\left(4s^2 - 1\right)\left(2s + 3\right)}\frac{a^{2s}}{R_o^{\,2s}} \text{ ....(23)};$$

and for points inside the sphere [i.e. $R_o < \alpha$]

$$\sum_m \frac{\gamma m}{R} = \frac{3\gamma M'}{2a}\left(1 - \frac{1}{3}\frac{R_o^{\,2}}{a^2}\right) \text{ ....(24)},$$

$$A^{\circ} = \frac{\gamma M'}{2c^2 a}\left\{R_0^{\ 2} - \frac{1}{10}\frac{R_0^{\ 4}}{a^2} + \frac{3}{2}a^2\right\} \text{..................} (25).$$

The Associated Space. We now introduce new coordinates (X_1, X_2, X_3, X_4) which are not in general pure spatio-temporal coordinates (unless the attracting bodies be at rest in the *x*-space), but are closely associated with the Cartesian coordinates (x_1, x_2, x_3, x_4). We write

$$X_\mu = x_\mu - \frac{1}{\omega_\mu^{\ 2}}\frac{\partial A^{\circ}}{\partial x_\mu}, \quad [\mu = 1, 2, 3, 4] \text{..............} (26).$$

We can then by an easy transformation deduce

$$\left.\begin{aligned} dJ^2 &= \left(c^2 + 6\Phi^{\circ}\right) dX_4^{\ 2} - \left(1 + \frac{2}{c^2}\Psi_4^{\ \circ}\right)\sum_\mu{}' dX_\mu^{\ 2} \\ &\quad -8\sum_\rho \Psi_\rho^{\ \circ} dX_\rho dX_4 + 4\sum_\rho \frac{\partial B^{\circ}}{\partial X_\rho} dX_\rho dX_4 \end{aligned}\right\} \text{.........} (27).$$

It is to be noticed that (X_1, X_2, X_3, X_4) are transformed to (Y_1, Y_2, Y_3, Y_4) cogrediently with the transformation of (x_1, x_2, x_3, x_4) to (y_1, y_2, y_3, y_4). Hence we can conceive that (X_1, X_2, X_3, X_4) and (Y_1, Y_2, Y_3, Y_4) are two sets of rectangular coordinates to an event-particle in an 'Associated Space-Time Continuum.'

Then corresponding to a path in the *x*-space traversed with velocity ($\dot{x}_1$, $\dot{x}_2$, $\dot{x}_3$,), there is a path in the associate space traversed with velocity ($\dot{X}_1$, $\dot{X}_2$, $\dot{X}_3$), where

$$\dot{X}_\mu = \frac{dX_\mu}{dX_4}, \quad [\mu = 1, 2, 3, 4].$$

Also we write

$$\left.\begin{aligned} V_M^{\ 2} &= \sum_\mu{}' \dot{X}_\mu^{\ 2} \\ \text{and} \qquad dS^2 &= \sum_\mu{}' dX_\mu^{\ 2} \end{aligned}\right\} \text{..............................} (28).$$

Now put

$$\left.\begin{aligned} L &= \tfrac{1}{2} v_M^2 + \sum_m \frac{\gamma m}{R} \\ K &= \tfrac{1}{2} v_M^2 - \sum_m \frac{\gamma m}{R} \\ \Delta &= \left(1 + \frac{2}{c^2} \sum_m \frac{\gamma m}{R}\right) \Big/ \left(1 - \frac{2}{c^2} L\right)^{\frac{1}{2}} \end{aligned}\right\} \quad \ldots\ldots\ldots\ldots(29).$$

The equations of motion [cf. equation (6), Chapter v, and equation (27) of this chapter] now become [for $\mu = 1,2,3$]

$$\left.\begin{aligned} &\frac{d}{dX_4}\left[M\Delta \dot{X}_\mu\right] + 4M \sum_\rho{}' \Psi^{\circ}{}_{\mu\rho} \dot{X}_\rho \\ &\quad = M\Delta \left(1 + \frac{2}{c^2} K\right) \frac{\partial}{\partial X_\mu}\left(4\Psi_4{}^{\circ} - 3\Phi^{\circ}\right) \\ &\qquad -4M \frac{\partial \Psi_u{}^{\circ}}{\partial X_4} + E \sum_\rho F_{\mu\rho}^{(x)} \dot{X}_\rho \end{aligned}\right\} \quad \ldots\ldots(30),$$

where

$$\Psi^{\circ}{}_{\mu\rho} = \frac{\partial \Psi_\mu{}^{\circ}}{\partial x_\rho} - \frac{\partial \Psi_\rho{}^{\circ}}{\partial x_\mu} \quad \ldots\ldots\ldots\ldots\ldots\ldots(31),$$

and (since $\| F_\mu \|$ is a covariant tensor)

$$F_\rho^{(X)} = F_\rho^{(x)} + \sum_a \frac{1}{c^2 \omega_a^2} F_a^{(x)} \frac{\partial^2 A^{\circ}}{\partial x_\rho \, \partial x_a} \quad \ldots\ldots\ldots\ldots(32).$$

We note that throughout the small terms we can neglect the distinction between the true and the associate continua.

CHAPTER VII

PERMANENT GRAVITATIONAL FIELDS

WHEN the attracting masses $(m_1, m_2, \ldots)$ are permanently at rest in the x-space, we obtain those particular cases of gravitational action for which Einstein's general equations of condition have been solved.

We now have

$$\left.\begin{aligned} \Phi &= \Psi_4 = \Phi^{\circ} = \Psi_4^{\circ} = \sum_m \frac{\gamma m}{R} \\ \Psi_\mu &= \Psi_\mu^{\circ} = 0, \quad [\mu = 1, 2, 3] \\ X_4 &= x_4 \end{aligned}\right\} \quad \ldots\ldots\ldots\ldots\ldots\ldots\ldots\ldots (1).$$

Hence [cf. equation (27) of Chapter VI]

$$dJ^2 = \left(c^2 - 2\Psi_4\right) dX_4^2 - \left(1 + \frac{2}{c^2}\Psi_4\right) \sum_\mu{}' \, dX_\mu^2 + 4dB \,.\, dX_4 \quad \ldots\ldots\ldots (1{\cdot}1),$$

and [cf. equation (12) of Chapter VI]

$$dJ^2 = \left(c^2 - 2\Psi_4\right) dx_4^2 - \left(1 + \frac{2}{c^2}\Psi_4\right) \sum_\mu{}' \, dx_\mu^2 + 2\sum_\mu{}' \sum_\nu{}' \frac{\partial^2 A}{\partial x_\mu \partial x_\nu} dx_\mu dx_\nu + 4dB \,.\, dx_4 \quad \ldots\ldots\ldots (1{\cdot}2).$$

Thus the equations of motion become

$$\left.\begin{aligned}&\frac{d}{dX_4}\left[M\Delta\dot{X}_\mu\right]=M\Delta\left(1+\frac{2}{c^2}K\right)\frac{\partial}{\partial X_\mu}\Psi_4\\&+E\sum_\rho F_{\mu\rho}^{(X)}\dot{X}_\rho,\ \left[\mu=1,2,3\right]\end{aligned}\right\}\ldots\ldots\ldots\ldots\ldots(2).$$

CHAPTER VIII

APPARENT MASS AND THE SPECTRAL SHIFT

IN the first place consider the vibration of some internal part of a molecule. Let M be its mass and V its undisturbed velocity. Then in the absence of gravitation

$$\Delta = \left(1 - \frac{V^2}{c^2}\right)^{-\frac{1}{2}} \qquad (1),$$

and in the presence of gravitation

$$\Delta = \left(1 + \frac{3}{c^2}\Psi_4{}^{\circ}\right)\left(1 - \frac{V^2}{c^2}\right)^{-\frac{1}{2}} \qquad (2).$$

But in either case $M\Delta$ is the effective mass. Accordingly, assuming that the electromagnetic forces which bind together the molecule are unaltered by the presence of the gravitational field, the period of vibration is lengthened in this field from

$$T \text{ to } T + \delta T,$$

where

$$\delta T/T = \frac{3}{2c^2}\Psi_4{}^{\circ} \qquad (3).$$

But the electromagnetic forces will be affected by the field. Accordingly, it requires some knowledge of the structure of the molecule to be certain what the shift (if any) of the spectral lines should be.

For example, assuming the electromagnetic laws considered under Chapters X and XIII below and assuming that the cohesive forces of an atom depend on the statical distribution of electric charges, the presence of the gravitational potential will (on the average) change any such cohesive force from F to

$$\left(1+\frac{2}{3c^2}\Psi_4\right) F_4,$$

according to the formula (10.1) of Chapter XIII below. Thus the shift would now become

$$\delta T/T = \frac{7}{6c^2}\Psi_4^{\,0} \qquad \text{.................................}(4).$$

The whole question is discussed in detail in Chapters XIII, XIV, and XV below.

CHAPTER IX

PLANETARY MOTION

LET the sun be the only gravitating body and let it be permanently at rest in the *x-space* at the spatial origin of the coordinate-system (x_1, x_2, x_3). The corresponding polar coordinates are (r_1, r_2, r_3). Then

$$\Psi_4 = \gamma m / r_1, \quad A = \gamma m r_1 \quad \text{.............................(1).}$$

Hence
$$X_\mu = x_\mu\left(1 - \frac{\gamma m}{c^2 r_1}\right), [\mu = 1, 2, 3] \quad \text{.......................(2).}$$

We then put
$$R_1 = \sqrt{\sum_\mu{}' X_\mu{}^2} = r_1\left(1 - \frac{\gamma m}{c^2 r_1}\right) \quad \text{.........................(3),}$$

and evidently (R_1, r_2, r_3) are the polar coordinates in the associate space of the point (X_1, X_2, X_3).

Consider uniplanar motion of a planet in the plane

$$r_3 = 0.$$

The equations of motion become

$$\frac{d}{dx_4}(\Delta \dot{R}_1) - \Delta R_1 \dot{r}_2^2 = \Delta\left(1 + \frac{2K}{c^2}\right)\frac{\partial \Psi_4}{\partial R_1} \quad \text{................(4),}$$

and
$$\frac{d}{dx_4}\left(\Delta R_1^2 \dot{r}_2\right) = 0 \quad \text{...............................(5).}$$

Thus
$$\Delta R_1^2 \dot{r}_2 = h \quad \text{.......................................(6).}$$

Then transferring to r_2 as independent variable and putting

$$u = 1/R_1 \quad \text{...(7)},$$

we find

$$\frac{d^2u}{dr_2^{\,2}} + u = \frac{\Delta^2}{h^2}\left(1 + \frac{2K}{c^2}\right)\frac{d\Psi_4}{du} \quad \text{........................(8)}.$$

But, to our approximation,

$$\Psi_4 = \gamma m u - \frac{\gamma^2 m^2 u^2}{c^2} \quad \text{..............................(9)}.$$

Thus the equation becomes

$$\frac{d^2u}{dr_2^{\,2}} + \left(1 - \frac{6\gamma^2 m^2}{c^2h^2}\right)u = \frac{\gamma m}{h^2}\left(1 + \frac{4}{c^2}K\right) \quad \text{................(10)}.$$

Now K is the constant energy of the orbit on the supposition that c is infinite. Hence, to our approximation, $c^2 K$ is constant.

We now put

$$u = l^{-1}\left(1 + e\cos\theta\right) \quad \text{............................(11)},$$

where

$$\theta = \left(1 - k\right)r_2 + \alpha \quad \text{...............................(12)},$$

and immediately deduce

$$l^{-1} = \frac{\gamma m}{h^2}\left(1 - \frac{2\gamma^2 m^2}{c^2h^2} + \frac{4}{c^2}K\right) \quad \text{........................(13)}$$

and

$$k = \frac{3\gamma^2 m^2}{c^2h^2} = \frac{3\gamma m}{c^2 l} \quad \text{..............................(14)}.$$

This value for k is Einstein's result.

The path of the planet in the x-space is

$$\frac{1}{r_1} + \frac{\gamma m}{c^2 r_1^{\,2}} = l^{-1}\left(1 + e\cos\theta\right) \quad \text{......................(15)}.$$

It is to be noticed that we have not assumed that e is small.

It is evident that Kepler's second law receives a slight modification, since

$$R^2 \dot{r}_2 = h\Delta^{-1} \quad \text{....................................(16)}.$$

CHAPTER X

ELECTROMAGNETIC EQUATIONS

WE have to consider a modification of the Maxwell-Lorentz Equations which will exhibit an influence of the gravitational field on the electromagnetic field.

The electric and magnetic forces in the x-space are expressed by the skew symmetric tensor

$$\| F_{\mu\nu} \|,$$

where

$$F_{\mu\nu}^{(x)} = \frac{\partial F_{\mu}^{(x)}}{\partial x_{\nu}} - \frac{\partial F_{\nu}^{(x)}}{\partial x_{\mu}} \quad \text{..............................(1).}$$

Hence it follows that, if

$$\lambda \mp \mu \mp \nu,$$

then

$$\frac{\partial F_{\mu\nu}}{\partial x_{\lambda}} + \frac{\partial F_{\nu\lambda}}{\partial x_{\mu}} + \frac{\partial F_{\lambda\mu}}{\partial x_{\nu}} = 0, \quad [\lambda, \mu, \nu = 1, 2, 3, 4] \text{............(2).}$$

Thus one set of four out of the Maxwell-Lorentz equations is identically satisfied. We now choose the remaining set of four in a form which exhibits a gravitational influence. Let $\| J^{\mu\nu} \|$ denote the contravariant tensor conjugate to $\| J_{\mu\nu} \|$ defined by

$$\left.\begin{aligned} \sum_{\rho} J_{(x)}^{\mu\rho} J_{\nu\rho}^{(x)} &= 0, \quad [\mu \mp \nu] \\ &= 1, \quad [\mu = \nu] \end{aligned}\right\} \text{..........................(3).}$$

We then define the skew contravariant tensor $\|(JF)^{\mu\nu}_{(x)}\|$ by

$$(JF)^{\mu\nu}_{(x)} = \sum_{\rho}\sum_{\sigma} J^{\rho\mu}_{(x)} J^{\sigma\nu}_{(x)} F^{(x)}_{\rho\sigma} \qquad \text{.........................(4).}$$

Then since $\|(JF)^{\mu\nu}_{(x)}\|$ is skew,

$$\left\| \sum_{\alpha} \frac{\partial}{\partial x_{\alpha}} (JF)^{\mu\alpha}_{(x)} \right\|$$

is a contravariant tensor.

Also let $\rho_{(x)}$ be the electric density in the x-space, so that

$$\left\| \rho_{(x)} \dot{x}_{\mu} \right\|$$

is the Contravariant Electric Motion Tensor. Then the second set of the remaining four of the Maxwell-Lorentz equations is

$$\sum_{\alpha} \frac{\partial}{\partial x_{\alpha}} (JF)^{\mu\alpha}_{(x)} = \frac{4\pi\rho_{(x)}}{c^2} \dot{x}_{u}, \quad [\mu = 1, 2, 3, 4] \qquad \text{..............(5).}$$

If the circumstances are such that the gravitational elements [i.e. $J^{(x)}_{\mu\nu}$]can be taken as constant throughout the region of the electromagnetic field, then [cf. eqn. (3) above] we can write these equations in the covariant form

$$\sum_{\alpha}\sum_{\sigma} J^{\sigma\alpha}_{(x)} \frac{\partial F^{(x)}_{\lambda\sigma}}{\partial x_{\alpha}} = \frac{4\pi\rho_{(x)}}{c^2} \sum_{\mu} J^{(x)}_{\lambda\mu} \dot{x}_{\mu}, \quad [\lambda = 1, 2, 3, 4] \qquad \text{........(6).}$$

CHAPTER XI

GRAVITATION AND LIGHT WAVES

THE wave-lengths of light waves are short compared with the linear dimensions of any region within which the gravitational elements vary. Also it is possible to assign regions such that the gravitational elements are constant within them and yet large enough to contain areas of wave-fronts of linear dimensions large compared to the wave-lengths. Also any lengths expressive of the curvatures of the wave-fronts may be assumed to be large compared to the linear dimensions of such regions, at least in the application considered below.

It follows that the characteristic equations for light waves in uncharged space are [cf. equation (6) of Chapter x]

$$\sum_{\alpha}\sum_{\sigma} J_{(x)}^{\sigma\alpha} \frac{\partial F_{\mu\sigma}^{(x)}}{\partial x_{\alpha}} = 0 \quad \text{..............................(1).}$$

In the corresponding coordinates of the associate continuum these equations become

$$\sum_{\alpha}\sum_{\sigma} J_{X}^{\sigma\alpha} \frac{\partial F_{\mu\sigma}^{(X)}}{\partial X_{\alpha}} = 0 \quad \text{..............................(2).}$$

Also with our assumption as to the constancy of the J's, $X\mu$ is a linear function of x_1, x_2, x_3, x_4. Hence a plane wave in the x-space is a plane wave in the associate X-space.

Now assume

$$F_{\mu}^{(X)} \propto e^{\iota\left(VX_4 - X_1\right)}, \ \left[\mu = 1, 2, 3, 4\right] \quad \text{....................(3).}$$

Then our conditions give

$$F^{(X)}_{23} = 0 \qquad\qquad (4).$$

$$\left.\begin{aligned} VJ^{44}_{(X)} - J^{41}_{(X)} &= \left(VJ^{14}_{(X)} - J^{11}_{(X)}\right)\Big/V \\ &= \left\{\left(J^{21}_{(X)} - VJ^{24}_{(X)}\right)F_2 \right. \\ &\quad \left. + \left(J^{31}_{(X)} - VJ^{34}_{(X)}\right)F_3\right\}\Big/\left(VF_1 + F_4\right) \end{aligned}\right\} \qquad (5).$$

We derive therefore a determination of $VF_1 + F_4$ as a multiple of the small quantity

$$\left\{\left(J^{21}_{(X)} - VJ^{24}_{(X)}\right)F_2 + \left(J^{31}_{(X)} - VJ^{34}_{(X)}\right)F_3\right\},$$

and the equation

$$V^2 J^{44}_{(X)} - 2VJ^{14}_{(X)} + VJ^{11}_{(X)} = 0 \qquad\qquad (6)$$

for the determination of V.

Assume that the gravitating bodies are permanently at rest in the x-space. Then [cf. equation (1·1) of Chapter VII and equation (3) of Chapter X]

$$\left.\begin{aligned} J^{11}_{(X)} &= -\left(1 - \frac{2}{c^2}\Psi_4\right) = -\frac{1}{c^2}J^{(X)}_{44} \\ J^{44}_{(X)} &= \frac{1}{c^2}\left(1 + \frac{2}{c^2}\Psi_4\right) = -\frac{1}{c^2}J^{(X)}_{11} \\ J^{14}_{(X)} &= \frac{1}{c^2}J^{(X)}_{14} \end{aligned}\right\} \qquad (7).$$

Also, measuring along the normal (in X-space) to the plane wave,

$$V = \frac{dX_1}{dX_4} \qquad\qquad (8).$$

Hence the equation for V becomes

$$dJ^2 = 0 \qquad\qquad (9).$$

Thus Einstein's assumption is proved for the normal advance of very short electromagnetic waves, such as the light waves, considered as advancing in associate space. This result holds for any short waves for which the radii of curvature of the wave-fronts are large compared to the wave-lengths. For then a small area of wave-front can be treated as plane.

Now consider a ray from a fixed point P to a fixed point Q in the x-space. By Huyghens' principle its course is given by making the time T to be stationary for small variations of the path between these points. Now by comparison with the associate space, since the gravitational field is permanent, X_4 and x_4 are identical. Thus

$$T = \int_P^Q \frac{ds}{v} = \int_P^Q \frac{dS}{V} \quad \text{(10)}.$$

But V is given by

$$c^2 - 2\Psi_4 - \left(1 + \frac{2}{c^2}\Psi_4\right)V^2 + 4\frac{dB}{dS}V = 0 \quad \text{(11)}.$$

Now V is nearly equal to c. Hence to our approximation

$$V = c\left\{1 - \frac{2}{c^2}\Psi_4 + \frac{2}{c}\frac{dB}{dS}\right\} \quad \text{(12)}.$$

Thus $$T = \frac{1}{c}\int_P^Q \left(1 + \frac{2}{c^2}\Psi_4\right) dS - \frac{2}{c^2}\left(B_Q - B_P\right) \quad \text{(13)}.$$

Thus, keeping P and Q fixed,

$$\delta T = \frac{1}{c}\delta \int_P^Q \left(1 + \frac{2}{c^2}\Psi_4\right) dS \quad \text{(14)}.$$

Thus the associate path of the ray in the associate space is obtained by assuming the associate space to be filled with a medium of refractive index

$$1 + \frac{2}{c^2}\Psi_4.$$

Also in the particular case when the gravitational field is due to the sun, the refractive index is

$$1 + \frac{2\gamma m}{c^2 R},$$

and since A is a function of r_1 only, the polar coordinates (r_1, r_2, r_3) in the x-space correspond to the polar coordinates (R, r_2, r_3) in the X-space. Hence the angle subtended at the sun by the two points at infinity on the ray in the x-space is equal to the analogous angle subtended at the sun by the two points at infinity on the associate ray in the X-space. Thus Einstein's result as to the deviation immediately follows.

Furthermore, it follows from the expression for T, that no modification of interference fringes can arise, due to the terms in dJ^2 involving B, by the use of an apparatus by which alternative rays for light, originating from the same source at P, are sent along alternative paths from P to Q, since

$$\left[T\right]_{\text{one path}}^{\text{other path}} = \left[\frac{1}{c}\int_P^Q \left(1 + \frac{2}{c^2}\Psi_4\right) dS\right]_{\text{one path}}^{\text{other path}} \quad \text{..........(15).}$$

If Ψ_4 be constant along the paths, this equation becomes

$$\left[T\right]_{\text{one path}}^{\text{other path}} = \frac{1}{c}\left(1 + \frac{2}{c^2}\Psi_4\right)\left[S\right]_{\text{one path}}^{\text{other path}} \quad \text{............(15·1).}$$

Now on the surface of the earth, if the axes of coordinates be fixed relatively to the surface and the axis of x_3 be vertically upwards, we have [cf. equations (22) and (26) of Chapter VI]

$$X_1 = x_1, \quad X_2 = x_2,$$

$$X_3 = x_3 - \frac{4}{5}\frac{ga^2}{c^2},$$

where g is the gravitational acceleration and a is the earth's radius.

It follows that $$S = s.$$

Accordingly, if the alternative geometrical paths of a divided ray be of equal geometrical length,

$$\left[T\right]_{\text{one path}}^{\text{other path}} = 0.$$

Thus in any experiment of the Michelson-Morley type, the earth's gravitational field will produce no modification of the interference fringes. The null result of the Michelson-Morley experiment is therefore fully explained.

CHAPTER XII

TEMPERATURE EFFECTS ON GRAVITATIONAL FORCES

ASSUME that the attracting body is at rest, except that its separate molecules have a velocity of agitation of which the mean square is $\bar{u}^2$. Let *Av* stand for 'Average value of.' Consider

$$AvdJ^2.$$

We use the formula of Chapter VI, equation (12), for dJ^2 and the values of the potentials given in equations (7) to (11) of that chapter. Then R refers to the permanent position of a molecule, neglecting its agitation due to temperature.

Write

$$\overline{\Psi} = \sum_m \frac{\gamma m}{R}, \quad \overline{A} = \sum_m \frac{\gamma m R}{c^2}, \quad \overline{B} = \sum_m \frac{\gamma m}{c} \log R, \quad \text{.............}(1),$$

and note that

$$Av \frac{\xi_0^2}{R^2} = \frac{1}{3} \frac{\overline{u}^2}{c^2} \quad \text{..................................}(2)$$

Then we easily find

$$\left. \begin{aligned} AvdJ^2 = &\left\{ c^2 - 2\left(1 + \frac{7}{6}\frac{\overline{u}^2}{c^2}\right)\overline{\Psi} \right\} dx_4^2 \\ &- \left(1 + \frac{2}{c^2}\overline{\Psi}\right) \sum_\mu{}' dx_\mu^2 + 2\sum_\mu{}'\sum_\nu{}' \frac{\partial^2 \overline{A}}{\partial x_\mu \partial x_\nu} dx_\mu\, dx_\nu \\ &+ 4\sum_\mu{}' \frac{\partial \overline{B}}{\partial x_\mu} dx_\mu\, dx_\nu \end{aligned} \right\} \text{.....}(3).$$

Thus the gravitational potential Ψ requires the coefficient

$$\left(1+\frac{7}{6}\frac{\overline{u}^2}{c^2}\right)$$

due to the temperature of the attracting body.

The coefficient due to the temperature of the attracted body is complicated by the change of apparent mass due to the velocity of agitation and by the possible effect of this velocity on the electromagnetic forces. Accordingly the special circumstances must be known before any calculation can be applied.

Returning to the consideration of the correction for the temperature of the attracting body, let G_1 and $u_1{}^2$ be its gravitational attraction at a given point and the mean square of its molecular velocities when its absolute temperature is T_1, and G_0 and $u_0{}^2$ be analogous quantities when its absolute temperature is T_0. Then

$$u_1^2=\alpha T_1,\quad u_0^2=\alpha T_0 \quad\text{.............................}(4),$$

where α is some constant depending on the physical constitution of the body. For example, for water vapour at 0°,

$$u_0^2=5\times 10^9\left[\text{cm.}^2/\text{sec.}^2\right],\ T_0=273^\circ \quad\text{................}(5).$$

Hence

$$\alpha=1{\cdot}83\times 10^7\ (\text{approx.}) \quad\text{...........................}(6).$$

Also

$$\frac{G_1-G_0}{G_0}=\frac{7}{6}\frac{u_1^2-u_0^2}{c^2}=\frac{7}{6}\frac{\alpha\left(T_1-T_0\right)}{c^2} \quad\text{..................}(7).$$

Hence, putting

$$\alpha=\alpha'\times 10^7 \quad\text{.....................................}(8).$$

we find

$$\frac{G_1-G_0}{G_0}=1{\cdot}29\times 10^{-14}\times\alpha'\left(T_1-T_0\right)\text{..................}(9).$$

Thus the temperature correction due to ordinary molecular velocities is inappreciable. It may be the case however that we should consider the velocities within each molecule, particularly within the nucleus of an atom. It may well be that such velocities attain to a sensible fraction of the velocity of light. The temperature correction might then be appreciable.

CHAPTER XIII

THE ELECTROSTATIC POTENTIAL AND SPECTRAL SHIFT

SINCE the gravitational field is permanent [cf. equation (1·2) of Chapter VII]

$$\left.\begin{aligned} dJ^2 = \left(c^2 - 2\Psi_4\right) dx_4^2 - \left(1 + \frac{2}{c^2}\Psi_4\right) \sum_\mu{}' dx_\mu^2 \\ +2\sum_\mu{}'\sum_\nu{}' \frac{\partial^2 A}{\partial x_\mu \partial x_\nu} dx_\mu\, dx_\nu + 4dB\,.\,dx_4 \end{aligned}\right\} \text{..........}(1)$$

We also assume (as in Chapter XI) that in the region considered the gravitational field is practically uniform, so that the spatial rate of variation of $g_{\mu\nu}$ is negligible. This amounts to neglecting gravitational accelerations but retaining gravitational potentials or quasi-potentials.

We write $$A_{\mu\nu} \equiv \frac{\partial^2 A}{\partial x_\mu\, \partial x_\nu} \text{....................................}(2),$$

so that Ψ_4 and $A_{\mu\nu}$ $[\mu, \nu = 1, 2, 3]$ are constants throughout the region considered.

Neglecting terms involving c^4 as a factor, we find

$$J^{\mu\mu}_{(x)} = -\left\{1 - 2\left(\frac{1}{c^2}\Psi_4 - A_{\mu\mu}\right)\right\}, [\mu = 1, 2, 3] \text{..............}(3),$$

$$J^{\mu\nu}_{(x)} = -\,2A_{\mu\nu}, \qquad [\mu \neq \nu \neq 4]\text{..........}(4),$$

$$J^{44}_{(x)} = \frac{1}{c^2}\left(1 + \frac{2}{c^2}\Psi_4\right) \quad \text{.................................}(5),$$

$$J^{\mu 4}_{(x)} = \frac{2}{c^2}\frac{\partial B}{\partial x_\mu}\left[= 0\right], \qquad [\mu \neq 4] \quad \text{...............}(6).$$

Hence putting $\lambda = 4$ in equation (6) of Chapter x, we find that in a steady electromagnetic field and a permanent gravitational field, the equation for the electrostatic potential F_4 becomes

$$\nabla^2 F_4 + 2\sum_\mu{}' \sum_\nu{}' A_{\mu\nu} \frac{\partial^2 F_4}{\partial x_\mu \partial x_\nu} = -4\pi\rho_{(x)} \quad \text{..................}(7).$$

Hence the solution for a point-charge e at the origin is

$$F_4 = \frac{e}{r_1}\left\{1 + \sum_\mu{}' \sum_\nu{}' \frac{A_{\mu\nu} x_\mu x_\nu}{r_1^2}\right\} \quad \text{..........................}(8),$$

where r_1 is the distance of the point (x_1, x_2, x_3).

Now consider a number of molecules in the region each forming an isolated electrical system. Let each molecule have an axis-system (x_1, x_2, x_3) at the centre of its nucleus, and let each such axis-system have the same relation to its electrical configuration as any other such axis-system has. But the molecules are orientated in every possible manner with respect to the gravitating field.

Hence if e be the charge at the centre of the nucleus which can be conceived as keeping the molecule together, and $-e'$ be the charge of any part of the molecule whose vibration is being considered, it follows that on the average the cohesive radial force is

$$\frac{ee'}{r_1^2}\left\{1 + \tfrac{1}{3}\left(A_{11} + A_{22} + A_{33}\right)\right\} \quad \text{.........................}(9).$$

But

$$\nabla^2 A = \frac{2}{c^2}\Psi_4 \quad \text{..................................}(10).$$

Hence the average cohesive force on that element of the molecule is

$$\frac{ee'}{r_1^2}\left(1 + \frac{2}{3c^2}\Psi_4\right) \quad \text{............................}(10\cdot 1).$$

But the apparent mass of the element due to the gravitational potential is

$$M\left(1+\frac{3}{c^2}\Psi_4\right),$$

where M is its apparent mass in the absence of the field. Hence, if T be the period of vibration of the element in the absence of the gravitational field and $T+\delta_1 T$ in its presence,

$$\frac{\delta_1 T}{T}=\frac{7}{6c^2}\Psi_4 \quad \text{..................................(11).}$$

Einstein's formula for the shift of the spectral lines is

$$\frac{\delta_1 T}{T}=\frac{1}{c^2}\Psi_4 \quad \text{..................................(12).}$$

For observational purposes the two formulae are indistinguishable.

CHAPTER XIV

THE LIMB EFFECT

LET $$x_\mu = \alpha_\mu r_1, \quad [\mu = 1, 2, 3] \quad \ldots\ldots(1).$$

Then [cf. equation (8) of Chapter XIII]

$$F_4 = \frac{e}{r_1}\left\{1 + \sum_\mu{}' \sum_\nu{}' A_{\mu\nu}\alpha_\mu\alpha_\nu\right\} \ldots\ldots(2).$$

Here $(\alpha_1, \alpha_2, \alpha_3)$ are the direction-cosines of the vector from the origin to (x_1, x_2, x_3). Thus the radial force is

$$-\frac{\partial F_4}{\partial r_1} = \frac{e}{r_1^2}\left\{1 + \sum_\mu{}' \sum_\nu{}' A_{\mu\nu}\alpha_\mu\alpha_\nu\right\} \ldots\ldots(3).$$

Now consider the internal vibration of a molecule which radiates light of period T (in a non-gravitational field) as capable of being represented as the vibration of a variable electric Hertzian doublet with this period.

Let $(\alpha_1, \alpha_2, \alpha_3)$ be the direction-cosines of the axis of the doublet. Then owing to the gravitational field the electric force which controls the vibration of the doublet is changed by the presence of the factor

$$\left\{1 + \sum_\mu{}' \sum_\nu{}' A_{\mu\nu}\alpha_\mu\alpha_\nu\right\}.$$

Let T become $T+\delta' T$ owing to the joint effect of this factor and of the change in the apparent mass of the electrons forming the doublet (due to the gravitational field). Then

$$\frac{\delta' T}{T} = \left\{ \frac{3\Psi_4}{2c^2} - \tfrac{1}{2} \sum_\mu{}' \sum_\nu{}' A_{\mu\nu}\alpha_\mu\alpha_\nu \right\} \quad \text{.......................(4).}$$

Let there be a large number of electrons forming the atmosphere of a star (say, the sun). Let the observer be at a great distance along the axis of x_1. Put

$$\alpha_1 = \cos \alpha_1, \quad \alpha_2 = \sin \alpha_1 \cos \alpha_2, \quad \alpha_3 = \sin \alpha_1 \sin \alpha_2 \quad \text{.......(5).}$$

Now doublets radiate light unequally in different directions. The intensity (measured by the energy radiated) varies as the square of the sine of the colatitude of the direction, the latitude being reckoned from the equatorial plane of the doublet. Thus the intensity of the light from the doublet in direction $(\alpha_1, \alpha_2, \alpha_3)$ sent to the observer varies as $\sin^2 \alpha_1$. Also the average change of period $(\delta' T)$ of the light sent to the observer with colatitude α_1 (as reckoned from the equatorial planes of the doublets) is given by

$$\frac{1}{2\pi}\int_0^{2\pi} \frac{\delta' T}{T}\, da_2,$$

i.e. by

$$\left.\begin{aligned} Av\frac{\delta' T}{T} &= \frac{1}{2}\left\{\frac{3\Psi_4}{c^2} - A_{11}\cos^2\alpha_1 - \tfrac{1}{2}\left(A_{22} + A_{33}\right)\sin^2\alpha_1\right\} \\ &= \frac{1}{2c^2}\Psi_4\left(3 - \sin^2\alpha_1\right) \\ &\qquad\qquad + \tfrac{1}{2}A_{11}\left(\tfrac{1}{2}\sin^2\alpha_1 - \cos^2\alpha_1\right) \end{aligned}\right\} \quad \text{.........(6).}$$

Now the light from the molecules for which α_1 is nearly 90° will be the brightest, both because of the factor $\sin^2 \alpha_1$ in the intensity, and because the equatorial belt of angular space of breadth $d\alpha_1$ is greater than the belts of the same angular breadth as α_1 approaches zero. Hence the shift of the spectral lines will approach that given by taking $\alpha_1 = \pi/2$. This conclusion is reinforced by the discussion of the next chapter on permanent directions of vibration—at least so far as relates to the centre or the edge of the sun's disc. Thus

$$Av\frac{\delta' T}{T} = \frac{1}{c^2}\Psi_4 + \tfrac{1}{4}A_{11} \quad\text{..............................(7).}$$

Now let α be the radius of the sun, and let the centre of the sun be the point

$$\left(-\alpha\cos\beta_1,\ -\alpha\cos\beta_2,\ -\alpha\cos\beta_3\right),$$

so that at the point of the sun's surface from which the light is taken the direction-cosines of the upward vertical are

$$\left(\cos\beta_1,\ \cos\beta_2,\ \cos\beta_3\right).$$

Let R be the distance from the molecule at (x_1, x_2, x_3) to the centre of the sun. Then after differentiating we can put α for R, and zero for x_1, x_2, x_3.

Then $$A = \frac{\gamma M}{c^2 R}\left(R^2 + \eta\alpha^2\right) \quad\text{..............................(8),}$$

where $$0 < \eta < \tfrac{1}{5} \quad\text{..............................(9).}$$

It follows from equation (22) of Chapter VI that η would be exactly $\frac{1}{5}$ if the sun were homogeneous. But it is probably considerably smaller.

Then $$A_{11} = \frac{\partial^2 A}{\partial x_1^2} = \left\{2\eta + \left(1 - 3\eta\right)\sin^2\beta_1\right\}\frac{\Psi_4}{c^2} \quad\text{...............(10).}$$

Thus $$\frac{\delta' T}{T} = \frac{1}{c^2}\Psi_4\left\{1 + \tfrac{1}{2}\eta + \tfrac{1}{4}\left(1 - 3\eta\right)\sin^2\beta_1\right\} \quad\text{.............(11).}$$

This formula exhibits a Limb Effect. For if the light comes from the centre of the sun, then

and $$\left.\begin{aligned}&\beta_1 = 0,\\ &\frac{\delta'' T}{T} = \frac{1}{c^2}\Psi_4\left(1 + \tfrac{1}{2}\eta\right)\end{aligned}\right\} \quad\text{..............................(12).}$$

and if the light comes from the edge of the disc, then

$$\beta_1 = \pi/2,$$

and $$\frac{\delta''' T}{T} = \frac{1}{c^2}\Psi_4 \frac{1}{4}(5-\eta) \quad \text{.........................(13).}$$

Hence, as we proceed from the sun's centre to its rim, there is a shift of spectral lines towards the red, defined by

$$\frac{\delta^{iv} T}{T} = \frac{1}{c^2}\Psi_4 \frac{1}{4}(1-3\eta) \quad \text{.........................(14).}$$

Thus, if we take $\eta = \frac{1}{10}$

(which is probably not far from the truth), we find

$$\frac{\delta^{iv} T}{T} = \frac{7}{40c^2}\Psi_4 \quad \text{...............................(15).}$$

It is unnecessary to point out the roughness of the assumptions, particularly the conception of the molecule as a vibrating doublet emitting light. But the investigation does suffice to show that our general assumptions do require the existence of a limb effect of the same order and sign as that actually observed.

CHAPTER XV

PERMANENT DIRECTIONS OF VIBRATION AND THE DOUBLING EFFECT

CONSIDER a vibrating element of a molecule of charge—e', the charge of the central nucleus being e. Let (P_1, P_2, P_3) be the mechanical force on the element due to the electrostatic attraction of the nucleus. Then [cf. equation (8) of Chapter XIII]

$$\left.\begin{aligned} P_s = -\frac{ee'x_s}{r_1^3}\left\{1 + 3\sum_\mu{}'\sum_\nu{}' \frac{A_{\mu\nu}x_\mu x_\nu}{r^2}\right\} \\ + \frac{2ee'}{r_1^2}\sum_\mu{}' A_{\mu s}x_\mu, \ \left[s = 1, 2, 3\right] \end{aligned}\right\} \quad \text{.............(1).}$$

Let this element of the molecule be that element whose radial vibration in direction $(\alpha_1, \alpha_2, \alpha_3)$ constitutes the variable doublet to which the radiation of the light is due. This direction of vibration cannot be permanent unless the force (P_1, P_2, P_3) is in the direction $(\alpha_1, \alpha_2, \alpha_3)$, when

$$x_s = r_1\alpha_s, \ \left[s = 1, 2, 3\right] \quad \text{..............................(2).}$$

Hence for permanence we require

$$\sum_\mu{}' A_{\mu s}\alpha_\mu \propto \alpha_s, \quad \left[s = 1, 2, 3\right] \quad \text{.........................(3).}$$

Consider a molecule in the atmosphere of the sun as in the previous chapters. Then [cf. equations (8) and (10) of Chapter XIV]

$$\left.\begin{aligned} A_{\mu\nu} &= -(1-3\eta)\cos\beta_\mu \cos\beta_\nu . \frac{1}{c^2}\Psi_4, \; [\mu \neq \nu] \\ A_{\mu\mu} &= \left[-(1-3\eta)\cos^2\beta_\mu + (1-\eta)\right]\frac{1}{c^2}\Psi_4 \end{aligned}\right\} \quad \text{...........(4)}.$$

Hence

$$\sum_\mu{}' A_{\mu s}\alpha_\mu = -\frac{1-3\eta}{c^2}\Psi_4 \cos\beta_s \sum_\mu{}' \alpha_\mu \cos\beta_\mu + \frac{1-\eta}{c^2}\Psi_4 \alpha_s \quad \text{.......(5)}.$$

Hence

$$\left.\begin{aligned} &\textit{either} \quad \sum_\mu{}' \alpha_\mu \cos\beta_\mu = 0 \\ &\textit{or} \quad \cos\beta_s = \alpha_s, \; [s = 1, 2, 3] \end{aligned}\right\} \quad \text{..........................(6)}.$$

Thus a permanent direction of vibration must be either normal or tangential to the gravitational level surface.

Accordingly, in the gaseous mass of molecules forming the atmosphere of the sun there will be an excess of molecules with their vibrations either normal to the level surface or in one of the directions tangential to the level surface.

First consider the vibrations normal to the level surface, and as in the previous chapter let the observer be on the axis of x_1 at a great distance. Then for these vibrations we should put

$$\alpha_s = \cos\beta_s, [s = 1, 2, 3] \quad \text{..............................(7)}.$$

Thus [cf. equation (4) of Chapter XIV]

$$\begin{aligned} \frac{\delta_1 T}{T} &= \left\{ \frac{3\Psi_4}{2c^2} - \tfrac{1}{2}\sum_\mu{}'\sum_\nu{}' A_{\mu\nu}\cos\beta_\mu \cos\beta_\nu \right\} \\ &= \frac{1}{2c^2}\Psi_4 . \left[3 + (1-3\eta)\left\{ \sum_\mu{}' \cos^2\beta_\mu \right\}^2 - (1-\eta)\sum_\mu{}' \cos^2\beta_\mu \right] \\ &= \frac{3-2\eta}{2c^2}\Psi_4 \quad \text{..(8)}. \end{aligned}$$

Thus these molecules yield a constant shift of the spectral lines all over the sun's disc. But the intensity of the light due to them varies as $\sin^2\beta_1$. Accordingly, they should yield faint lines from the centre of the disc and comparatively strong lines from its edge.

Secondly, consider molecules vibrating tangentially to the sun's gravitational level surface. No generality is lost by taking the axes of x_2 and x_3, so that the sun's diameter through the point of the disc considered is in the plane $x_1\, x_2$. In this case

$$\beta_3 = \frac{\pi}{2}, \quad \beta_1 + \beta_2 = \frac{\pi}{2} \quad \text{..............................(9)}.$$

Also the level surface at the point contains the axis of x_3.

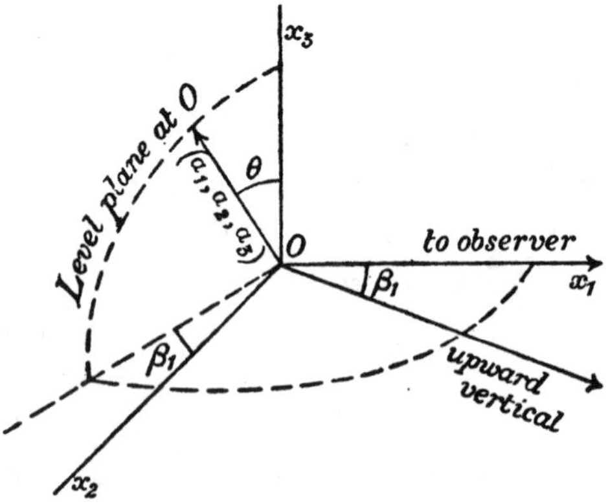

Then we can put

$$\left.\begin{aligned} \alpha_1 &= -\sin\theta \sin\beta_1 = -\sin\theta\cos\beta_2 \\ \alpha_2 &= \sin\theta\cos\beta_1 \\ \alpha_3 &= \cos\theta \end{aligned}\right\} \quad \text{..................(10)}$$

Hence equation (4) of Chapter XIV becomes

$$\begin{aligned} \frac{\delta_2 T}{T} &= \frac{3}{2c^2}\Psi_4 \\ &\quad -\tfrac{1}{2}\sin^2\theta\left(A_{11}\cos^2\beta_2 - 2A_{12}\cos\beta_1\cos\beta_2 + A_{22}\cos^2\beta_1\right) \\ &\qquad -\tfrac{1}{2}A_{33}\cos^2\theta - \sin\theta\cos\theta\left(A_{23}\cos\beta_1 - A_{31}\cos\beta_2\right). \end{aligned}$$

By the use of equations (4) of this chapter, this reduces to

$$\frac{\delta_2 T}{T} = \frac{2+\eta}{2c^2}\Psi_4 \quad \text{.................................(11).}$$

To consider the comparative brightness of light from these molecules taken at different points [i.e. for different values of β_1] note that [cf. equations (10) of this chapter]

$$\sin^2\alpha_1 = 1 - \alpha_1^{\,2} = 1 - \sin^2\theta\,\sin^2\beta_1 \quad \text{..................(12).}$$

Hence the total light from this type of molecule is brightest at the centre of the disc [$\beta_1 = 0$], since for every value of θ

$$\sin^2\alpha_1 = 1 \quad \text{....................................(13).}$$

The brightness falls off as we pass towards the edge of the disc, and finally at the edge $\left(\beta_1 = \frac{\pi}{2}\right)$

$$\sin^2\alpha_1 = \cos^2\theta \quad \text{................................(14).}$$

Also the average value for $\sin^2\alpha_1$ at any point is

$$1 - \tfrac{1}{2}\sin^2\beta_1 \quad \text{....................................(15).}$$

It will also be noticed that the larger angular area of an equatorial belt of angular breadth $2\delta\alpha_1$ over a polar cap of angular diameter $2\delta\alpha_1$ gives the tangential molecules another advantage in brightness over those of the former type vibrating normally to the level surface.

To sum up the discussion on the shift of the spectral lines contained in this and the two preceding chapters: The molecules in the sun's atmosphere can be sorted into three groups, (i) a group of molecules uniformly pointing in all directions, (ii) a group of molecules pointing normally to the gravitational level, and (iii) a group of molecules pointing tangentially to the gravitational level. A molecule is said to point in the direction of the equivalent electric doublet whose variation generates the light waves.

The relative brightness of the light from these three groups changes as we pass from the centre to the edge of the sun's disc. It is constant for group (i), it varies as $\sin^2\beta_1$ for group (ii), and it varies as $(1-\frac{1}{2}\sin^2\beta_1)$ for group (iii).

The spectral shift for group (i) is on one hypothesis [cf. Chapter XIII]

$$\frac{7}{6c^2}\Psi_4,$$

and on another hypothesis [cf. equation (11) of Chapter XIV] it approximates to

$$\frac{1}{c^2}\Psi_4\left\{1+\tfrac{1}{2}\eta+\tfrac{1}{4}(1-3\eta)\sin^2\beta_1\right\}$$

where η is probably not greater than $\frac{1}{10}$.

The spectral shift for group (ii) is

$$\frac{3-2\eta}{2c^2}\Psi_4.$$

The spectral shift for group (iii) is

$$\frac{2+\eta}{2c^2}\Psi_4.$$

Accordingly, in light derived from the sun, or a star, or a nebula in (1) a general shift of the spectral lines to the red may be expected; (2) since groups (ii) and (iii) change in relative importance as we pass from the centre to the edge of the disc, and since the shift due to group (i) also changes, so a shift of spectral lines towards the red (the limb effect) may be expected; (3) in the case of the sun or a nebula some evidence of a doubling or even a trebling of the spectral lines may be expected.

It is also to be noticed that the number of vibrations of a doublet emitting light from the visible spectrum during a mean free path of the molecule is of the order 10^4, even allowing for the fact that the velocity of the molecule is largely due to a high temperature. Accordingly, within each mean free path there is time for the vibrations to settle down into one of their permanent directions.

Finally, we note that when it shall be possible to measure with reasonable accuracy the spectral shifts of light from the stars and the nebulae, we obtain a numerical determination for the mass divided by the radius of the body concerned. Hence if either the mass or the radius be known, the other can be found.

CHAPTER XVI

STEADY ELECTROMAGNETIC FIELDS

THE equations (6) of Chapter X for a steady electromagnetic field become

$$\sum_{\alpha}{}' \sum_{\sigma}{}' J_{(x)}^{\sigma\alpha} \frac{\partial F_{\lambda\sigma}}{\partial x_\alpha} + \sum_{\alpha}{}' J_{(x)}^{4\alpha} \frac{\partial F_{\lambda 4}}{\partial x_\alpha}$$

$$= \frac{4\pi\rho_{(x)}}{c^2}\left[J_{\lambda 4}^{(x)} + \sum_{\mu}{}' J_{\lambda\mu}^{(x)} \dot{x}_\mu\right], \left[\lambda = 1, 2, 3\right] \text{........(1)}.$$

Consider a region where there is no current and let $F_{\mu\nu}{}^{\circ}$ $[\mu, \nu = 1, 2, 3, 4]$ be the value of $F_{\mu\nu}$ which is the first approximation when the gravitational influence is neglected.

We use equations (3) to (6) of Chapter XIII and put

$$4\pi\rho_{(x)} = \frac{\partial F_{14}{}^{\circ}}{\partial x_1} + \frac{\partial F_{24}{}^{\circ}}{\partial x_2} + \frac{\partial F_{34}{}^{\circ}}{\partial x_3} \text{ ....(2)}.$$

We also take Ω as the magnetic potential for the approximate magnetic force $(cF_{23}{}^{\circ}, cF_{31}{}^{\circ}, cF_{12}{}^{\circ})$, so that

$$\left(cF_{23}{}^{\circ},\ cF_{31}{}^{\circ},\ cF_{12}{}^{\circ},\right) = -\text{grad } \Omega \text{........................(3)}.$$

The equations to determine the magnetic force

$$\left(cF_{23},\ cF_{31},\ cF_{12},\right)$$

now become

$$\operatorname{curl}\left(S_1,\ S_2,\ S_3,\right) = 0 \quad \ldots\ldots\ldots\ldots\ldots\ldots(4),$$

where

$$S_1 = cF_{23} - 2\sum_{\nu}{}' A_{1\nu}\frac{\partial\Omega}{\partial x_\nu} + \frac{2}{c}\left(\frac{\partial B}{\partial x_3}\frac{\partial F_4{}^{\circ}}{\partial x_2} - \frac{\partial B}{\partial x_2}\frac{\partial F_4{}^{\circ}}{\partial x_3}\right) \ldots\ldots(5),$$

with analogous meanings for S_2 and S_3.

Hence the second approximation gives

$$\left.\begin{aligned} cF_{23} = cF_{23}{}^{\circ} - 2\left\{A_{11}cF_{23}{}^{\circ} + A_{12}cF_{31}{}^{\circ} + A_{13}cF_{12}{}^{\circ}\right\} \\ -\frac{2}{c}\left\{\frac{\partial B}{\partial x_2}F_{34}{}^{\circ} - \frac{\partial B}{\partial x_3}F_{24}{}^{\circ}\right\}\end{aligned}\right\}\ldots\ldots\ldots(6),$$

$$\left.\begin{aligned} cF_{31} = cF_{31}{}^{\circ} - 2\left\{A_{21}cF_{23}{}^{\circ} + A_{22}cF_{31}{}^{\circ} + A_{23}cF_{12}{}^{\circ}\right\} \\ -\frac{2}{c}\left\{\frac{\partial B}{\partial x_3}F_{14}{}^{\circ} - \frac{\partial B}{\partial x_1}F_{34}{}^{\circ}\right\}\end{aligned}\right\}\ldots\ldots\ldots(7),$$

$$\left.\begin{aligned} cF_{12} = cF_{12}{}^{\circ} - 2\left\{A_{31}cF_{23}{}^{\circ} + A_{32}cF_{31}{}^{\circ} + A_{33}cF_{12}{}^{\circ}\right\} \\ -\frac{2}{c}\left\{\frac{\partial B}{\partial x_1}F_{24}{}^{\circ} - \frac{\partial B}{\partial x_2}F_{14}{}^{\circ}\right\}\end{aligned}\right\}\ldots\ldots\ldots(8).$$

In the first place we note that a steady electric force ($F_{14}{}^{\circ}$, $F_{24}{}^{\circ}$, $F_{34}{}^{\circ}$) in a permanent gravitational field produces the magnetic force

$$\frac{2}{c}\left[\left(F_{14}{}^{\circ}, F_{24}{}^{\circ}, F_{34}{}^{\circ}\right).\text{grad } B\right],$$

where $[R.R']$ stands for the vector product of the two vectors R and R'.

Accordingly, the magnetic force is perpendicular to the electric force which produces it and to the vector *grad B*.

Consider a field on the surface of the earth. Let α be the earth's radius and let the axis of x_1 be the upward vertical. Then at the origin (which is on the earth's surface)

$$\text{grad } B = \left(\frac{\epsilon}{c}\Psi_4, 0, 0\right) \quad\text{..............................(9),}$$

where [cf. equation (23) of Chapter VI]

$$0{\cdot}76 \ldots < \epsilon < 1.$$

Here ϵ would attain its lower limit if the earth were uniform throughout. We shall assume

$$\epsilon = 0{\cdot}88 \quad\text{..(10)}$$

as a sufficient approximation in the actual circumstances.

Hence the magnetic force produced by the electric force

$$(F \cos\alpha, \quad F \sin\alpha, 0)$$

is

$$\frac{2\epsilon}{c^2}\Psi_4 F. (0, 0, -\sin\alpha),$$

i.e. is the horizontal force

$$\frac{2\epsilon g\alpha}{c^2} F \sin\alpha,$$

perpendicular to the vertical plane containing the electric force and proportional to the sine of the angle which the electric force makes with the vertical. Here g denotes the ordinary gravitational acceleration.

Accordingly, a given electric force produces the greatest magnetic effect when it is horizontal. But in any case the magnetic force produced is extremely small, being about

$$1{\cdot}2 \times 10^{-9} \times F \sin\alpha \text{ (gausses)},$$

where F is the measure of the electric force in electrostatic units.

The corresponding effect on the surface of the sun would be about

$$3{\cdot}8 \times 10^{-5} \times F \sin\alpha \text{ (gausses)}.$$

This effect is the only effect I have found which depends on the existence of B. Accordingly, an experiment of sufficient accuracy to detect the magnetic force, if it exists, would be of great interest as forming a crucial experiment to test the formula for dJ^2 here adopted.

A steady magnetic field is also modified by the presence of the gravitational field.

For example, consider a current I (electromagnetic measure) along the axis of x_1, and let R be the distance of (x_1, x_2, x_3) from this axis. Then

$$\left.\begin{aligned} cF_{23}^{0} &= 0 \\ cF_{31}^{0} &= -\frac{2Ix_3}{R^2} \\ cF_{12}^{0} &= \frac{2Ix_2}{R^2} \end{aligned}\right\} \quad \text{.................................(11).}$$

Let the wire (i.e. the axis of x_1) make an angle β_1 with the upward vertical, and let the axis of x_2 lie in the vertical plane through the wire. Also let the plane through the wire and the point (x_1, x_2, x_3) make an angle ϕ with this vertical plane through the wire. Then we find [cf. equations (4) of Chapter xv]

$$\left.\begin{aligned} A_{12} &= -(1-3\eta)\sin\beta_1\cos\beta_1 . \frac{1}{c^2}\Psi_4 \\ A_{13} &= 0 \end{aligned}\right\} \quad \text{.................(12).}$$

Hence [cf. equation (6) above]

$$cF_{23} = -\frac{(1-3\eta)\Psi_4}{c^2} . \frac{2I}{R} . \cos\phi \sin 2\beta_1 \quad \text{.................(13).}$$

Thus there is a small magnetic force parallel to the wire which is equal to

$$-\frac{(1-3\eta)g\alpha}{c^2}\cos\phi\sin 2\beta_1 \times \frac{2I}{R}$$

at distance R from the wire.

This force vanishes if the wire be vertical or horizontal and is greatest when the wire is inclined at an angle of 45° to the vertical. Also it is greatest in the vertical plane through the wire, and vanishes in the plane through the wire perpendicular to this vertical plane.

Thus its greatest value at a distance R from the wire

$\left[\beta_1 = \frac{\pi}{4}, \text{ and } \phi = 0 \text{ or } \pi\right]$ is

$$\frac{(1-3\eta)g\alpha}{c^2} \times \frac{2I}{R} \text{ (gausses)}.$$

Hence, taking $\eta = \frac{1}{10}$, its greatest value at distance R from the wire is about

$$\tfrac{1}{2} \times 10^{-9} \times \frac{2I}{R} \text{ (gausses)}.$$

CHAPTER XVII

THE MOON'S MOTION

ASTRONOMICAL tables, which depend on observations made at all times of the year, must finally register spatio-temporal elements in terms of the space-time which is the rest-system of the sun. We must therefore distinguish between relative motion and difference motion in respect to a given space-tune. Thus the relative motion of the moon with respect to the earth is the motion of the moon in the earth's rest-system at the moment of observation. But the difference motion of the moon from the earth in the sun's rest-space is the vector excess of the motion of the moon over that of the earth reckoned in the sun's rest-space. On the classical theory of a unique space and unique time difference motion and relative motion were identical. We have to treat them as distinct with distinct formulae. It is evident that astronomical tables for the moon concern the difference motion of the moon from the earth with respect to the sun's rest-system of space-time.

Let (x_1, x_2, x_3, x_4) be the coordinates of the moon at the time x_4 with respect to rectangular coordinate axes in the sun's rest-space with the sun as origin. Let the contemporary [i.e. at the same sun-time x_4] coordinates of the earth be (q_1, q_2, q_3). We now take the earth as a moving origin in the x-space and obtain the difference coordinates for the moon, referred to moving axes parallel to the fixed axes (y_1, y_2, y_3), where

$$y_\mu = x_\mu - q_\mu \quad [\mu = 1, 2, 3] \quad \text{..........................(1);}$$

also the difference coordinates for the sun referred to the earth as moving origin are $(-q_1, -q_2, -q_3)$, where the three positions in the sun's rest-space for the sun, earth and moon are contemporary at the time x_4.

The difference velocity of the moon from the earth is therefore $(\dot{y}_1, \dot{y}_2, \dot{y}_3)$.

Let v be the magnitude of the velocity of the moon in the sun's rest-space, and let U be the velocity of the earth in the same space. Also let V be the magnitude of the difference velocity of the moon from the earth. Then

$$v^2 = V^2 + U^2 + 2\sum_\mu{}' \dot{y}_\mu \dot{q}_\mu \quad \text{..........................(2).}$$

Also we write

$$\left.\begin{aligned} \xi &= \frac{1}{c}\sum_\mu{}' \left(x_\mu - q_\mu\right)\dot{q}_\mu = \frac{1}{c}\sum_\mu{}' y_\mu \dot{q}_\mu \\ \eta &= \frac{1}{c^2}\sum_\mu{}' y_\mu \ddot{q}_\mu \end{aligned}\right\} \quad \text{....................(3).}$$

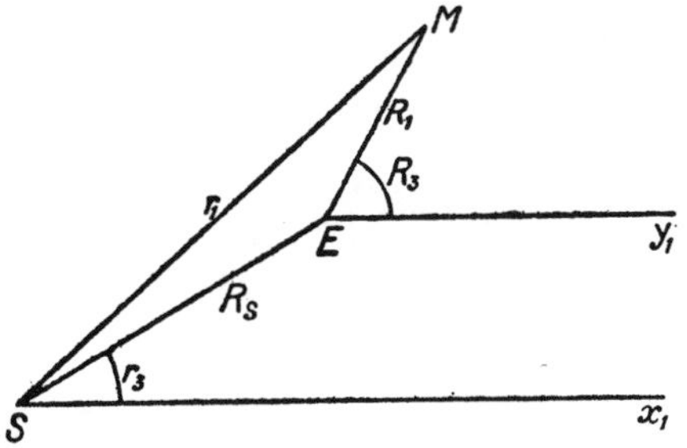

Assume a uniplanar motion in the plane of the ecliptic, so that

$$x_3 = 0, \; q_3 = 0, \; y_3 = 0 \quad \text{............................(4).}$$

Let (R_s, r_3) be the polar coordinates of the earth relatively to the sun, and let (R_1, R_3) be the polar difference-coordinates of the moon from the earth. Thus

$$\left.\begin{aligned} q_1 &= R_s \cos r_3, \quad q_2 = R_s \sin r_3 \\ y_1 &= R_1 \cos R_3, \quad y_2 = R_1 \sin R_3 \end{aligned}\right\} \qquad (5).$$

Also let r_1 be the distance of the moon from the sun, so that

$$r_1^{\,2} = R_1^{\,2} + R_s^{\,2} + 2R_1 R_s \cos \rho_3 \qquad (6),$$

where

$$\rho_3 = R_3 - r_3 \qquad (7).$$

We consider R_s and r_3 to be given functions of the time x_4, and take R_1 and R_3 as the coordinates to be determined in terms of the time by means of the equations of motion. Two propositions (*A* and *B*) are easily proved, of which particular cases are important: *Prop.* (*A*)

$$\frac{d}{dx_4}\frac{\partial}{\partial \dot{R}_1}\left(r_1^{\,n}\dot{r}_1\right) - \frac{\partial}{\partial R_1}\left(r_1^{\,n}\dot{r}_1\right) = 0 \qquad (8).$$

The important particular ease of this proposition is

$$\frac{d}{dx_4}\frac{\partial}{\partial \dot{R}_1}\frac{\dot{r}_1}{r_1} - \frac{\partial}{\partial R_1}\frac{\dot{r}_1}{r_1} = 0 \qquad (9).$$

Prop. (B)

$$\frac{d}{dx_4}\frac{\partial}{\partial \dot{R}_3}\left(r_1^{\,n}\dot{r}_1\right) - \frac{\partial}{\partial R_3}\left(r_1^{\,n}\dot{r}_1\right) = 0 \qquad (10).$$

The important particular case is

$$\frac{d}{dx_4}\frac{\partial}{\partial \dot{R}_3}\frac{\dot{r}_1}{r_1} - \frac{\partial}{\partial R_3}\frac{\dot{r}_1}{r_1} = 0 \qquad (11).$$

Let (p_1, p_2, p_3, p_4) be the coordinates of the antecedent position of the earth causally correlated to the moon's position at time x_4, referred to the fixed axes origin at the sun.

Let r_e be the distance between the moon and this antecedent position. Also put

$$\left.\begin{aligned} \xi_e &= \frac{1}{c}\sum_\mu{}' \left(x_\mu - p_\mu\right)\dot{p}_\mu \\ \Omega_e &= \left(1 - \frac{1}{c^2}\sum_\mu{}' \dot{p}_\mu{}^2\right)^{-\frac{1}{2}} \\ \Omega &= \left(1 - \frac{U^2}{c^2}\right)^{-\frac{1}{2}} \end{aligned}\right\} \text{..................(12).}$$

Then the potential impetus of the moon's route through the space-time manifold is [cf. equations (l) and (3) of Chapter VI] determined by

$$\left.\begin{aligned} dJ^2 = \left(c^2 - v^2\right)dx_4^2 - \frac{2\gamma m_s}{r_1}\left(1 - \frac{\dot{r}_1}{c}\right)^2 dx_4^2 \\ - \frac{2\gamma m_e}{\Omega_e^3\left(r_e - \xi_e\right)}\left(1 - \frac{d\dot{r}_e}{c\,dx_4}\right)^2 dx_4^2 \end{aligned}\right\} \text{.......(13),}$$

where *ms* and *me* are the masses of the sun and earth respectively.

We now put

$$L = \tfrac{1}{2}v^2 + \frac{\gamma m_s}{r_1} + \frac{\gamma m_e}{R_1}$$

$$+ \text{ terms due to planetary attraction ....(14).}$$

Thus
$$\Gamma = \left(1 - \frac{2}{c^2}L\right)^{\frac{1}{2}} \text{ ....(15).}$$

Also
$$\frac{1}{r_e} = \frac{1}{R_1}\left\{1 - \frac{\xi}{R_1} + \frac{1}{2}\frac{\xi^2}{R_1^2} - \frac{1}{2c^2}U^2 + \tfrac{1}{2}\eta\right\} \text{..............(16).}$$

$$\xi_e = \xi + \frac{1}{c^2}R_1U^2 - R_1\eta \text{..........................(17).}$$

Hence

$$\left.\begin{aligned} \tfrac{1}{2}c^2\Gamma^2 = \tfrac{1}{2}c^2 - L + H_e - \frac{\gamma m_s}{c^2}\frac{\dot{r}_1{}^2}{r_1} \\ + \frac{2\gamma m_s \dot{r}_1}{cr_1} + \frac{2\gamma m_e \dot{R}_1}{cR_1} \end{aligned}\right\} \text{..........(18).}$$

where

$$H_e = \frac{\gamma m_e}{R_1}\left[\frac{5}{2}\eta + \frac{1}{2}\frac{\xi^2}{R_1^2} + \frac{U^2 - \dot{R}_1^2}{c^2} + \frac{2}{c^2}\sum_\mu{}' \dot{y}_\mu \dot{q}_\mu\right] \quad \text{........(19).}$$

We put

$$\left.\begin{aligned} w_1 &= \dot{R}_s \cos\rho_3 + R_s\dot{r}_3 \sin\rho_3 \\ w_2 &= \dot{R}_s\dot{r}_3 \cos\rho_3 - \dot{R}_s \sin\rho_3 \\ \alpha_1 &= \left(\ddot{R}_s - R_s\dot{r}_3^2\right)\cos\rho_3 + \left(R_s\ddot{r}_3 + 2R_s\dot{r}_3\right)\sin\rho_3 \\ \alpha_2 &= \left(R_s\ddot{r}_3 + 2R_s\dot{r}_3\right)\cos\rho_3 - \left(\ddot{R}_s - R_s\dot{r}_3^2\right)\sin\rho_3 \end{aligned}\right\} \quad \text{....................(20).}$$

Then w_1, w_2, α_1, α_2 are functions of R_3 and x_4 only. Also

$$\left.\begin{aligned} &\xi = \frac{1}{c}R_1 w_1, \ \eta = \frac{1}{c^2}R_1\alpha_1 \\ &\sum_\mu{}' \dot{y}_\mu \dot{q}_\mu = \dot{R}_1 w_1 + R_1\dot{R}_3 w_2 \end{aligned}\right\} \quad \text{........................(21).}$$

Thus

$$H_e = \frac{\gamma m_e}{c^2 R_1}\left[\tfrac{5}{2}R_1\alpha_1 + \tfrac{1}{2}w_1^2 + 2\dot{R}_1 w_1 + 2R_1\dot{R}_3 w_2 + U^2 - \dot{R}_1^2\right] \quad \text{............(22).}$$

There are now [cf. equation (6) of Chapter v] two equations of motion of the type

$$\frac{d}{dx_4}\left\{\frac{1}{\Gamma}\frac{\partial}{\partial \dot{R}_\mu}\left(-\tfrac{1}{2}c^2\Gamma^2\right)\right\} - \frac{1}{\Gamma}\frac{\partial}{\partial R_\mu}\left(-\tfrac{1}{2}c^2\Gamma^2\right) = 0, \ [\mu = 1, 3] \quad \text{...........(23).}$$

It follows from the special cases of Props. (*A*) and (*B*), that the two terms involving c^{-1} as a factor disappear in both equations. Hence these equations both take the form [$\mu = 1, 3$],

$$\left.\begin{aligned} \frac{d}{dx_4}\frac{\partial L}{\partial \dot{R}_\mu} - \frac{\partial L}{\partial R_\mu} &= \frac{d}{dx_4}\frac{\partial}{\partial \dot{R}_\mu}\left(H_e - \frac{m_s\dot{r}_1^2}{c^2 r_1}\right) \\ &\quad - \frac{\partial}{\partial R_\mu}\left(H_e - \frac{m_s\dot{r}_1^2}{c^2 r_1}\right) - \frac{1}{c^2}\frac{dL}{dx_4}\frac{\partial L}{\partial \dot{R}_\mu} \end{aligned}\right\} \quad \text{.........(24).}$$

The terms on the left-hand side of these equations are those introduced by ordinary lunar theory; the terms on the right-hand side are the new small corrections introduced by the formulae of relativity of the form here adopted. I have not succeeded in eliciting any terms which, in the present state of the Lunar Tables, can be made the subject of comparison with observation. The investigation will therefore not be pursued further.

PART III

ELEMENTARY THEORY OF TENSORS

CHAPTER XVIII

FUNDAMENTAL NOTIONS

1. **Coordinates.** The mutual relations to each other of event-particles can be determined by characterising each event-particle by four measurements of four assigned types respectively. These four measurements are called the coordinates of the event-particle; and the four types of measurement must be such that (i) each type assigns to each event-particle one and only one coordinate of that type, and (ii) each set of four coordinates (as ordered in that assignment to types) characterises one and only one event-particle. Four given types of measurement with these properties are called a coordinate-system.

A coordinate-system will be called 'pure' if one of the coordinates be the time of some given space-time system 'x' and the other three coordinates be spatial-quantities of the space of the same system 'x.' A coordinate-system which is not pure is called 'mixed.' If (u_1, u_2, u_3, u_4) be the coordinates of an event-particle in a pure system, then it will be adopted as a convention that (u_1, u_2, u_3) represent the spatial coordinates of a point in the space of the space-time system to which the coordinates refer and u_4 represents the time in the same space-time system. Thus the event-particle (u_1, u_2, u_3, u_4) happens at the time u_4 and at the point (u_1, u_2, u_3) in the corresponding space.

If (u_1, u_2, u_3, u_4) and (x_1, x_2, x_3, x_4) be the coordinates of the same event-particles according to different coordinate-systems, the u-system and the x-system respectively, then there will be four equations of transformation

$$u_\mu = f_\mu\left(x_1, x_2, x_3, x_4,\right), \left[\mu = 1, 2, 3, 4\right] \ldots\ldots\ldots\ldots\ldots\ldots (1).$$

These four equations can be solved so as to give

$$x_\mu = F_\mu\left(u_1, u_2, u_3, u_4,\right), \left[\mu = 1, 2, 3, 4\right] \ldots\ldots\ldots\ldots\ldots (1\cdot 1).$$

If both systems are pure in the same space-time system, then

$$u_4 = x_4.$$

and (u_1, u_2, u_3) and (x_1, x_2, x_3) are different spatial coordinates of the same point in the space of that space-time system.

2. **Scalar Characters and Invariant Expressions.** Consider the measurement of some physical quantity arising in the physical field at an event-particle, such as the gravitational potential according to some definite meaning of that term. Its measure (so far as the definition of meaning is kept unchanged) must be independent of coordinate-systems. But its law of distribution throughout the various event-particles of space-time will be expressible as a function of the coordinates of the event-particles under consideration.

Such a physical character is called a scalar quantity.

We must distinguish between a scalar quantity and an invariant formula expressing that quantity. When a formula in terms of coordinates of relevant event-particles is such that it gives the same value for the scalar quantity whatever coordinate-system be employed, it is called an invariant formula. There may also be formulae which are only invariant for a limited set of systems of coordinates, derivable one from the other by transformations forming a group (in the mathematical sense of that term). In this limited case we have 'group invariance.'

When we can conceive a scalar character in such a way that it has no special or peculiar relation of any sort to one coordinate-system of a group which it has not to any other system of that group, it follows that there must be some group-invariant formula for the scalar character which is limited to that group of systems of coordinates.

3. **Physical Characters of the First Order.** A scalar character is a character of zero order.

A character of event-particles is of the first order when—given any coordinate-system (u_1, u_2, u_3, u_4)—it is expressible by an array of four quantities (functions of the coordinates of the event-particles in question) such that each quantity is specially related to one of the types of coordinate measurement. These four quantities are called the 'components' of that character for that coordinate-system.

For example, let (u_1, u_2, u_3, u_4) be a pure coordinate-system, and let a region of the *u*-space be filled with a continuously moving substance. Let the motion of the substance at (u_1, u_2, u_3) at the time u_4 be represented by

$$\frac{du_\mu}{du_4}, \quad \left[\mu = 1, 2, 3, 4\right].$$

Thus the array $(\dot{u}_1, \dot{u}_2, \dot{u}_3, 1)$ represents a character of the first order which is descriptive of the motion of the fluid.

Again, let α be the gravitational potential at the event-particle (u_1, u_2, u_3, u_4). Then the gradient

$$\left(\frac{\partial\Phi}{\partial u_1}, \frac{\partial\Phi}{\partial u_2}, \frac{\partial\Phi}{\partial u_3}, \frac{\partial\Phi}{\partial u_4}\right)$$

represents a character of the first order.

Let (x_1, x_2, x_3, x_4) be the coordinates in the coordinate-system '*x*' of the same event-particle as denoted by (u_1, u_2, u_3, u_4). Then $(\dot{x}_1, \dot{x}_2, \dot{x}_3, 1)$ is a character of the first order descriptive of the motion by reference to the coordinate-system '*x*.' We are at once brought to the consideration of the relations to each other of these two distinct descriptions of the same fact of motion by means of $(\dot{u}_1, \dot{u}_2, \dot{u}_3, 1)$ and $(\dot{x}_1, \dot{x}_2, \dot{x}_3, 1)$ respectively.

The relations between the two will be peculiarly simple (and therefore important) if the components of one character (say the *u*-character) are expressible as linear functions of the components of the other character (the *x*-character), where the coefficients may be functions of the coordinates of the event-particles in question which are purely determined by the general relations of the two coordinate-systems in question and are independent of the particular values of the components. Thus if $(T_1^{(u)}, T_2^{(u)}, T_3^{(u)}, T_4^{(u)})$ be a first order description of some fact in coordinate-system '*u*,' and

$$\left(T_1^{(x)},\ T_2^{(x)},\ T_3^{(x)},\ T_4^{(x)}\right)$$

be a description of the same fact in coordinate-system '*x*,' the desired linear relation is

$$T_{(u)}^{\mu} = \sum_{\alpha} l_{\mu\alpha}\, T_{\alpha}^{(x)},\ \left[\mu = 1, 2, 3, 4\right] \cdots\cdots\cdots\cdots (2),$$

where the coefficients $l_{\mu\alpha}$ $[\mu, \alpha=1, 2, 3, 4]$ are expressible in terms of the equations of transformation between '*u*' and '*x*' without any reference to the particular values of $T_1^{(x)}, T_2^{(x)}, T_3^{(x)}, T_4^{(x)}$.

4. **Tensors of the First Order.** Furthermore, we pass from the two assigned coordinate-systems '*u*' and '*x*' to the consideration of a group of systems (as in the case of invariance), if the determination of the coefficients [i. e. $l_{\mu\alpha}$] can be fixed by a general rule which is identical for any two systems of the group.

A first order character as thus described in any coordinate-system of a group is called a 'Group-Tensor' for that group.

If the general rule for the formation of coefficients in the linear relation between the components hold for all pairs of coordinate-systems whatsoever, then the character as thus described in all coordinate-systems is called a 'General Tensor' or more simply a 'Tensor.' It is a tensor of the first order.

It is obvious that in the case of a group-tensor or a general tensor the rule for the formation of the coefficients in the linear equations giving the components of the character for one coordinate-system in terms of the components for another coordinate-system must be such that the transformations of components from one system to another form a group. For there is to be only one description of the character in each coordinate-system. Accordingly, if '*u*,' '*v*,' '*x*' be three coordinate-systems and '*T*' a tensor character, then the transformation of

$$\left(T_1^{(u)}, T_2^{(u)}, T_3^{(u)}, T_4^{(u)}\right) \text{ to } \left(T_1^{(v)}, T_2^{(v)}, T_3^{(v)}, T_4^{(v)}\right)$$

and then of

$$\left(T_1^{(v)}, T_2^{(v)}, T_3^{(v)}, T_4^{(v)}\right) \text{ to } \left(T_1^{(x)}, T_2^{(x)}, T_3^{(x)}, T_4^{(x)}\right)$$

must give the same components in system 'x' as the direct transformation from system 'u' to system 'x.'

In future we will write

$$\left\| T^{(u)} \right\|$$

for the array of the components of a character in system 'u.'

5. **Covariant and Contravariant First-Order Tensors.** A tensor may refer to many event-particles. Suppose that one of these with a peculiar definite relation to the character in question is picked out and termed the dominant event-particle of the character. Let $\| T^{(u)} \|$ be the tensor in system 'u' and $\| T^{(x)} \|$ be the tensor in system 'x,' and let (u_1, u_2, u_3, u_4) and (x_1, x_2, x_3, x_4) be the coordinates of the dominant event-particle of the character expressed by the tensor.

The tensor is 'covariant' if its components in any system 'u' are related to its components in any system 'x' by

$$T_\mu^{(u)} = \sum_\alpha T_\alpha^{(x)} \frac{\partial x_\alpha}{\partial u_\mu}, \; [\mu = 1, 2, 3, 4] \quad \text{......................(3).}$$

In the case of contravariant tensors it is convenient to adopt an alternative notation $(T_{(u)}^1, T_{(u)}^2, T_{(u)}^3, T_{(u)}^4)$ for the components in any system 'u,' shortened into $\| T_{(u)} \|$ when the whole array is to be mentioned.

Then a tensor is 'contravariant' if its components in any system 'u' are related to its components in any system 'x' by

$$T_{(u)}^\mu = \sum_\alpha l_{\mu\alpha} T_\alpha^{(x)}, \; [\mu = 1, 2, 3, 4] \quad \text{......................(4).}$$

It is easy to prove the 'group' property of the covariant and contravariant modes of transformation by the use of the equations

$$\sum_\rho \frac{\partial u_\mu}{\partial v_\rho}\frac{\partial v_\rho}{\partial x_\alpha} = \frac{\partial u_\mu}{\partial x_\alpha} \quad \text{..............................(5)},$$

$$\left.\begin{aligned}\sum_\rho \frac{\partial u_\mu}{\partial v_\rho}\frac{\partial v_\rho}{\partial u_\nu} &= 0, \ [\mu \neq \nu] \\ &= 1, \ [\mu = \nu]\end{aligned}\right\} \quad \text{...............(5·1)}.$$

If the tensor property is restricted to a group of systems of coordinates, we obtain covariant group-tensors or contravariant group-tensors as the case may be.

6. **Characters and Tensors of Higher Orders.** A physical character is of the *n*th order when, in any coordinate system '*u*,' it is expressible by an array of 4^n quantities (functions of the coordinates of the event-particles in question) so that each component of the array is specially related to one permutation of the types of coordinate measurement, the types being taken *n* together in each permutation and repetitions of type being allowed.

Thus a character of the 2nd order will require the array

$$T_{\mu\nu}^{(u)}, \ [\mu, \nu = 1, 2, 3, 4].$$

For example, those seven components (out of the whole sixteen) which involve the coordinate-type 'u_1' are

$$T_{11}^{(u)}, T_{12}^{(u)}, T_{21}^{(u)}, T_{13}^{(u)}, T_{31}^{(u)}, T_{14}^{(u)}, T_{41}^{(u)}.$$

A character of the 3rd order will involve 64 components, and of the 4th order 256 components.

The same general explanations, *mutatis mutandis,* apply as in the case of characters of the 1st order.

The covariant tensor transformation (for the 2nd order) is

$$T_{\mu\nu}^{(u)} = \sum_\alpha \sum_\beta T_{\alpha\beta}^{(x)} \frac{\partial x_\alpha}{\partial u_\mu}\frac{\partial x_\beta}{\partial u_\nu} \quad \text{..........................(6)},$$

and the contravariant tensor transformation (for the 2nd order) is

$$T_{(u)}^{\mu\nu} = \sum_{\alpha}\sum_{\beta} T_{(x)}^{\alpha\beta} \frac{\partial u_\mu}{\partial x_\alpha} \frac{\partial u_\nu}{\partial x_\beta} \quad \text{.........................(6·1)},$$

and analogously for characters of other orders.

But 'mixed' tensors now appear in which both covariant and contravariant qualities are involved.

For example, a mixed tensor of the 2nd order [represented by the notation $T_\mu^{\ v}(u)$ for coordinate-system 'u'] is transformed by the rule

$$T_\mu^{\ \nu}(u) = \sum_{\alpha}\sum_{\beta} T_\alpha^{\ \beta}(x) \frac{\partial x_\alpha}{\partial u_\mu} \frac{\partial u_\nu}{\partial x_\beta} \quad \text{.........................(7)},$$

and analogously for higher orders.

7. **Tensor-Invariance of Formulae.** The tensor description of a physical character must not be confused with the tensor-invariance for mathematical formulae. If the array $\|T_\mu^{(v)}\|$ be an array of formulae involving the u-coordinates (u_1, u_2, u_3, u_4) as arguments, then these formulae have tensor-covariance if $\|T_\alpha^{(x)}\|$, as obtained from $\|T_\mu^{(v)}\|$ by the covariant rule of transformation, are expressible by the same formulae involving (x_1, x_2, x_3, x_4) as $\|T_\mu^{(v)}\|$ are expressed by the use of (u_1, u_2, u_3, u_4). Also similarly for tensor-contra-variance.

Thus tensor-invariance (as this property will be termed) implies the persistence of the same formulae after transference from one coordinate-system to another by means of the appropriate tensor formula (covariant or contravariant).

For example, if A be any function of the position of the event-particle (u_1, u_2, u_3, u_4), then the array

$$\left\| \frac{\partial A}{\partial u_\mu} \right\|$$

has tensor-covariance. For

$$\frac{\partial A}{\partial u_\mu} = \sum_{\alpha} \frac{\partial A}{\partial x_\alpha} \frac{\partial x_\alpha}{\partial u_\mu} \quad \text{.................................(8)}.$$

Again let dN be any homogeneous rational integral function of du_1, du_2, du_3, du_4 of the first degree, and thus analogously expressible in any coordinate-system. Also write

$$N^{(u)} = 1 \Big/ \frac{dN}{du_4},$$

where $\dot{u}_1$, $\dot{u}_2$, $\dot{u}_3$ represent the definite velocity of a substance at (u_1, u_2, u_3) at time u_4. Then

$$\left(N^{(u)}\dot{u}_1, N^{(u)}\dot{u}_2, N^{(u)}\dot{u}_3, N^{(u)}\right) \qquad \text{......................(8·1)}$$

has tensor-contravariance, since

$$du_\mu = \sum_\alpha dx_\alpha \frac{\partial u_\mu}{\partial x_\alpha}.$$

It is evident that the formulae expressing a law of nature which is not known to have any particular relation to the coordinate-systems in question should have tensor-invariance.

CHAPTER XIX

ELEMENTARY PROPERTIES

8. **Test for Tensor Property.** If an array character possess the tensor property (covariant, contravariant, or mixed) for transference from one given coordinate-system to every other coordinate-system, then it possesses it in general, namely for transference from any system to any other system. For let 'u' be the given coordinate-system and let 'p' and 'q' be any other coordinate-systems. As an example consider the mixed tensor $\|S_\mu^\nu\|$. Then by hypothesis

$$S_\mu^\nu(p) = \sum_\alpha \sum_\beta S_\alpha^\beta(u) \frac{\partial u_\alpha}{\partial p_\mu} \frac{\partial p_\nu}{\partial p_\beta} \quad \text{..........................(9)},$$

$$S_\rho^\sigma(q) = \sum_\gamma \sum_\delta S_\gamma^\delta(u) \frac{\partial u_\gamma}{\partial q_\rho} \frac{\partial p_\sigma}{\partial u_\delta} \quad \text{......................(9·1)}.$$

Multiply equation (9·1) by $\dfrac{\partial q_\rho}{\partial u_\alpha} \dfrac{\partial u_\beta}{\partial q_\sigma}$ and sum for ρ and σ.

[Note that in future this type of operation will be described as

'operating with $\sum_\rho \sum_\sigma \dfrac{\partial q_\rho}{\partial u_\alpha} \dfrac{\partial u_\beta}{\partial q_\sigma} *$.']

Then [cf. equations (5) and (5·1)]

$$\sum_{\rho}\sum_{\sigma} S_\rho^\sigma(q)\frac{\partial q}{\partial u_\alpha}\frac{\partial u_\beta}{\partial q_\sigma} = \sum_{\gamma}\sum_{\delta} S_\gamma^\delta(u)\sum_{\rho}\frac{\partial q_\rho}{\partial u_\alpha}\frac{\partial u_\gamma}{\partial q_\rho}\sum_{\sigma}\frac{\partial q_\sigma}{\partial u_\delta}\frac{\partial u_\beta}{\partial q_\sigma}$$
$$= S_\alpha^\beta(u).$$

Hence substituting in equation (9) for $S_\alpha^\beta(u)$

$$S_\mu^\nu(p) = \sum_{\alpha}\sum_{\beta}\sum_{\rho}\sum_{\sigma} S_\rho^\sigma(q)\frac{\partial q_\rho}{\partial u_\alpha}\frac{\partial u_\alpha}{\partial p_\mu}\frac{\partial p_\nu}{\partial u_\beta}\frac{\partial u_\beta}{\partial q_\sigma}$$
$$= \sum_{\rho}\sum_{\sigma} S_\rho^\sigma(q)\sum_{\alpha}\frac{\partial q_\rho}{\partial u_\alpha}\frac{\partial u_\alpha}{\partial p_\mu}\sum_{\alpha}\frac{\partial p_\nu}{\partial u_\beta}\frac{\partial u_\beta}{\partial q_\sigma}$$
$$= \sum_{\rho}\sum_{\sigma} S_\rho^\sigma(q)\frac{\partial q_\rho}{\partial p_\mu}\frac{\partial p_\nu}{\partial p_\sigma}.$$

This proves the required property.

9. **Sum of Tensors.** If $\|S\|$ and $\|T\|$ are two tensors of the same order and type, then $\|S+T\|$ and $\|S-T\|$ are tensors of that same order and type. The proof is obvious.

Again, if every component of a tensor vanishes in one coordinate-system, the same property holds in every coordinate-system.

It therefore follows that if $\|S\|$ and $\|T\|$ are tensors of the same order and type and their corresponding-components are equal in one coordinate-system, they are equal in every coordinate-system. This is the principle of tensor-equations.

10. **Product of Tensors.** Let $\|S\|$ be a tensor of the with order and $\|T\|$ a tensor of the nth order. Form a new array of the $(m + n)$th order whose components are the products of any component of $\|S\|$ with any component of $\|T\|$. This new array is a tensor with the covariant and contravariant affixes of both tensors.

As an example, let $\|S_\mu\|$ be a covariant tensor of the 1st order and $\|T_\nu{}^\rho\|$ be a mixed tensor of the 2nd order. We have then to prove that $\|S_\mu T_\nu{}^\rho\|$ is a mixed tensor of the 3rd order for which μ and v are the covariant affixes and ρ is the contravariant affix.

For

$$S_\mu^{(u)} = \sum_\alpha S_\alpha^{(x)} \frac{\partial x_\alpha}{\partial u_\mu},$$

$$T_\nu^{\rho}(u) = \sum_\beta \sum_\gamma T_\beta^{\gamma}(x) \frac{\partial x_\beta}{\partial u_\nu} \frac{\partial u_\rho}{\partial x_\gamma}.$$

Hence $$S_\mu^{(u)} T_\nu^{\rho}(u) = \sum_\alpha \sum_\beta \sum_\gamma S_\alpha^{(x)} T_\beta^{\gamma}(x) \frac{\partial x_\alpha}{\partial u_\mu} \frac{\partial x_\beta}{\partial u_\nu} \frac{\partial u_\rho}{\partial x_\gamma}.$$

This proves the proposition. An analogous proof holds for any other types of tensors.

11. **Representation of a Tensor as a Sum of Products.**

Case (i). If $\| T_{\mu\nu} \|$ be any covariant tensor of the 2nd order, then four pairs of covariant tensors of the 1st order can be found, namely

$$\| A_\mu \| \text{ and } \| A_\mu' \|, \quad \| B_\mu \| \text{ and } \| B_\mu' \|,$$
$$\| C_\mu \| \text{ and } \| C_\mu' \|, \quad \| D_\mu \| \text{ and } \| D_\mu' \|,$$

such that

$$T_{\mu\nu} = A_\mu A_\nu' + B_\mu B_\nu' + C_\mu C_\nu' + D_\mu D_\nu' \quad \text{................} (10).$$

For by sections 9 and 10 the left-hand side is a tensor of the right order and type. Hence, again by section 9, we have only to choose A, A', etc., so that in one coordinate-system the components of the composite system are equal to the components of T.

Consider the coordinate-system 'u.' In this system let

$$A_\mu = T_{\mu 1}, \quad [\mu = 1, 2, 3, 4],$$
$$A_1' = 1,$$
$$B_1' = C_1' = D_1' = 0.$$

Then $$T_{\mu 1} = A_\mu A_1' + B_\mu B_1' + \ldots, \quad [\mu = 1, 2, 3, 4].$$

Also let

$$B_\mu = T_{\mu 2}, \quad [\mu = 1, 2, 3, 4],$$
$$B_2' = 1,$$
$$A_2' = C_2' = D_2' = 0.$$

Then $$T_{\mu 2} = A_\mu A_2{}' + B_\mu B_2{}' + \ldots, \ [\mu = 1, 2, 3, 4].$$

Also treat $T_{\mu 3}$ and $T_{\mu 4}$ in the same way, so that

and
$$C_\mu = T_{\mu 3}, \ [\mu = 1, 2, 3, 4]$$
$$D_\mu = T_{\mu 4}, \ [\ \ ,, \ \ ,, \ \ ].$$

Hence in the 'u' system

$$T_{\mu\nu} = A_\mu A_\nu{}' + B_\mu B_\nu{}' + \ldots + \ldots,$$

and hence the equality holds in every system.

This proof will hold equally well for contravariant or mixed terms of the 2nd order.

Case (ii). A tensor of any type and of the 3rd order can be exhibited as a sum of products of four pairs of tensors, one tensor in each pair being of the 2nd order and one of the 1st order.

For example, consider the mixed tensor $\|T^\rho_{\mu\nu}\|$.

We can find four pairs of tensors,

$$A_{\mu\nu} \text{ and } \bar{A}^\rho, \quad B_{\mu\nu} \text{ and } \bar{B}^\rho, \quad C_{\mu\nu} \text{ and } \bar{C}^\rho, \quad D_{\mu\nu} \text{ and } \bar{D}_\rho,$$

such that

$$T^\rho_{\mu\nu} = A_{\mu\nu}\bar{A}^\rho + B_{\mu\nu}\bar{B}^\rho + \ldots + \ldots \ \ldots\ldots\ldots\ldots\ldots\ldots(11).$$

For as before we have merely to obtain the equality in one coordinate-system. Now take in coordinate-system 'u,'

$$A_{\mu\nu} = T^1_{\mu\nu}, \ [\mu, \nu = 1, 2, 3, 4],$$
$$\bar{A}^1 = 1,$$
$$\bar{B}^1 = \bar{C}^1 = \bar{D}^1 = 0,$$
$$B_{\mu\nu} = T^2_{\mu\nu}, \ [\mu, \nu = 1, 2, 3, 4],$$
$$\bar{B}^2 = 1,$$
$$\bar{A}^2 = \bar{C}^2 = \bar{D}^2 = 0,$$
$$C_{\mu\nu} = T^3_{\mu\nu}, \ [\mu, \nu = 1, 2, 3, 4],$$

$$\overline{C}^3 = 1,$$

$$\overline{A}^3 = \overline{B}^3 = \overline{D}^3 = 0,$$

$$D_{\mu\nu} = T^4_{\mu\nu}, \; \left[\mu, \nu = 1, 2, 3, 4\right],$$

$$\overline{D}^4 = 1,$$

$$\overline{A}^4 = \overline{B}^4 = \overline{C}^4 = 0.$$

The theorem is now proved.

It is obvious that this mode of representation can be proved successively for tensors of any order or type.

CHAPTER XX

THE PROCESS OF RESTRICTION

12. **Definition of Restriction.** Let $S_{\dots\rho}^{\dots}(u)$ be a tensor of any order, with ρ as a covariant affix, and otherwise of any type. Let $T_{\dots}^{\dots\rho}(u)$ be a tensor of any order, with ρ as a contravariant affix and otherwise of any type. Then the array

$$\sum_\rho S_{\dots\rho}^{\dots}(u)\, T_{\dots}^{\dots\rho}(u)$$

will be proved to be a tensor. It will be called a 'restricted product' of the two tensors $\|S\|$ and $\|T\|$. The order of the restricted product will obviously be two less than the sum of the orders of $\|S\|$ and $\|T\|$.

In the proof we will take the two tensors $S_{\mu\rho}^{\nu}(u)$ and $T_{(u)}^{\lambda\rho}$, but it will be obvious that the steps of the proof are absolutely general.

We have
$$S_{\mu\rho}^{\nu}(v) = \sum_\alpha \sum_\beta \sum_\gamma S_{\alpha\beta}^{\gamma}(u) \frac{\partial u_\alpha}{\partial v_\mu} \frac{\partial u_\beta}{\partial v_\rho} \frac{\partial u_\nu}{\partial v_\gamma},$$

$$T_{(v)}^{\lambda\rho} = \sum_\delta \sum_\epsilon T_{(v)}^{\delta\epsilon} \frac{\partial v_\lambda}{\partial v_\delta} \frac{\partial v_\rho}{\partial u_\epsilon}.$$

Hence

$$\sum_\rho S_{\mu\rho}^{\nu}(v)\, T_{(v)}^{\lambda\rho} = \sum_\alpha \sum_\beta \sum_\gamma \sum_\delta \sum_\epsilon S_{\alpha\beta}^{\gamma}(u)\, T_{(v)}^{\delta\epsilon} \frac{\partial u_\alpha}{\partial v_\mu} \frac{\partial v_\nu}{\partial u_\gamma} \frac{\partial v_\lambda}{\partial u_\delta} \sum_\rho \frac{\partial u_\beta}{\partial v_\rho} \frac{\partial v_\rho}{\partial u_\epsilon}.$$

Hence [cf. equation (5·1) of Chapter XVIII]

$$\sum_{\rho} S^{\nu}_{\mu\rho}(v)\, T^{\lambda\rho}_{(v)} = \sum_{\alpha}\sum_{\gamma}\sum_{\delta}\{\sum_{\beta} S^{\gamma}_{\alpha\beta}(u)\, T^{\delta\beta}_{(u)}\}\frac{\partial u_{\alpha}}{\partial v_{\mu}}\frac{\partial v_{\nu}}{\partial u_{\gamma}}\frac{\partial v_{\lambda}}{\partial u_{\delta}} \quad \text{.....(11·1).}$$

This proves the required tensor property, and an analogous proof is obviously applicable to all analogous cases.

13. **Multiple Restriction.** The analogous process of restriction can be applied for two or more pairs of contrasted indices [i.e. one index covariant and the other contravariant]. The multiply restricted product thus obtained will still be a tensor. If there be n processes of restrictive summation and m be the sum of the orders of the two tensors $\| S \|$ and $\| T \|$, the order of the multiply restricted tensor will be $m-2n$.

To prove the tensor property, take as an example

$$\sum_{\sigma}\sum_{\rho} S^{\nu}_{\sigma\rho}(v)\, T^{\sigma\rho}_{(v)}.$$

Now in equation (11·1) above put

$$\lambda = \mu = \sigma,$$

and sum for σ. Then [cf. equation (5·1)]

$$\sum_{\sigma}\sum_{\rho} S^{\nu}_{\sigma\rho}(v)\, T^{\sigma\rho}_{(v)} = \sum_{\alpha}\sum_{\gamma}\sum_{\delta}\{\sum_{\beta} S^{\gamma}_{\alpha\beta}(u)\, T^{\delta\beta}_{(u)}\}\frac{\partial v_{\nu}}{\partial u_{\gamma}}\sum_{\sigma}\frac{\partial u_{\alpha}}{\partial v_{\sigma}}\frac{\partial v_{\sigma}}{\partial u_{\delta}}$$

$$= \sum_{\gamma}\{\sum_{\alpha}\sum_{\beta} S^{\gamma}_{\alpha\beta}(u)\, T^{\alpha\beta}_{(u)}\}\frac{\partial v_{\nu}}{\partial u_{\gamma}} \quad \text{..................(11.2).}$$

14. **Invariant Products.** If the two tensors subject to restriction are of the same order n, and there be n-fold restriction, the order of the restricted product is zero, so that it is an invariant scalar quantity. For example,

$$\sum_{\mu}\sum_{\nu} S^{(v)}_{\mu\nu}\, T^{\mu\nu}_{(v)} = \sum_{\alpha}\sum_{\beta} S^{(u)}_{\alpha\beta}\, T^{\alpha\beta}_{(u)} \quad \text{........................(12),}$$

and

$$\sum_{\mu} S^{(v)}_{\mu}\, T^{\mu}_{(v)} = \sum_{\alpha} S^{(u)}_{\alpha}\, T^{\alpha}_{(u)} \quad \text{....................(12.1),}$$

and

$$\sum_{\mu}\sum_{\nu} S^{\nu}_{\mu}(v)\, T^{\mu}_{\nu}(v) = \sum_{\alpha}\sum_{\beta} S^{\beta}_{\alpha}(u)\, T^{\alpha}_{\beta}(u) \quad \text{............(12.2).}$$

15. **The Tensor** $\|I\|$. Let $I\mu v$ be defined in reference to a given coordinate-system 'u' by

$$\left.\begin{aligned} I_\mu^{\ \nu} &= 0, \left[\mu \mp \nu\right] \\ &= 1, \left[\mu \mp \nu\right] \end{aligned}\right\} \quad \ldots\ldots\ldots\ldots\ldots\ldots\ldots\ldots\ldots (13).$$

Consider the mixed tensor $\|I_\mu^v(v)\|$ whose components in the system 'u' are equal to those of the array $\|I_\mu^v\|$. Then

$$\begin{aligned} I_\mu^{\ \nu}(v) &= \sum_\alpha \sum_\beta I_\alpha^{\ \beta} \frac{\partial u_\alpha}{\partial v_\mu} \frac{\partial v_\nu}{\partial u_\beta} \\ &= \sum_\alpha \frac{\partial u_\alpha}{\partial v_\mu} \frac{\partial v_\nu}{\partial u_\alpha} \\ &= I_\mu^{\ \nu} \quad \ldots\ldots\ldots\ldots\ldots\ldots\ldots\ldots\ldots\ldots (13.1). \end{aligned}$$

Hence I_μ^v has the same relation to every coordinate-system as that which it has to coordinate-system 'u.'

16. **Restriction of a Single Mixed Tensor.** It follows from the theory of restriction that

$$\left\| \sum_\rho \sum_\sigma T_{\mu\rho}^{\sigma}(u)\, I_\sigma^{\rho} \right\|$$

is a tensor, i.e.

$$\left\| \sum_\rho T_{\mu\rho}^{\rho}(u) \right\|$$

is a tensor. Accordingly the mixed tensor of the 3rd order

$$\left\| T_{\mu\rho}^{\sigma}(u) \right\|$$

has been restricted into a tensor of the 1st order. This proof is obviously quite general. For example, if $\|T_\mu^{\ \nu}\|$ be a mixed tensor of the 2nd order, then

$$\sum_\mu T_\mu^{\ \mu}(u)$$

is invariant.

17. Argument from Products [Restricted or Unrestricted] to the Tensor Property.

This argument is best shown by a series of examples:

Case (i). If $\| T_\mu \|$ be an array of components, of the 1st order, defined for every coordinate-system, and if, whatever 1st-order contravariant tensor $\| S^\mu \|$ may be, we have the invariance of

$$\sum_\rho T_\rho^{(u)} S^\rho_{(u)},$$

then $\| T_\mu \|$ is a covariant tensor. For by hypothesis

$$\sum_\rho T_\rho^{(v)} S^\rho_{(v)} = \sum_\alpha T_\alpha^{(u)} S^\alpha_{(u)}$$

and
$$S^\rho_{(v)} = \sum_\alpha S^\alpha_{(u)} \frac{\partial v_\rho}{\partial u_\alpha}.$$

Hence
$$\sum_\alpha \left| T_\alpha^{(u)} - \sum_\rho T_\rho^{(v)} \frac{\partial v_\rho}{\partial u_\alpha} \right| S^\alpha_{(u)} = 0.$$

But the tensor $\| S^\alpha \|$ is arbitrary. Now make four successive choices

$$S_{(u)}{}^1 = 1, \quad S_{(u)}{}^2 = 0, \quad S_{(u)}{}^3 = 0, \quad S_{(u)}{}^4 = 0,$$
and
$$S_{(u)}{}^1 = 0, \quad S_{(u)}{}^2 = 1, \quad S_{(u)}{}^3 = 0, \quad S_{(u)}{}^4 = 0,$$
and
$$S_{(u)}{}^1 = 0, \quad S_{(u)}{}^2 = 0, \quad S_{(u)}{}^3 = 1, \quad S_{(u)}{}^4 = 0,$$
and
$$S_{(u)}{}^1 = 0, \quad S_{(u)}{}^2 = 0, \quad S_{(u)}{}^3 = 0, \quad S_{(u)}{}^4 = 1,$$

and substitute successively in the above equation. We obtain

$$T_\alpha^{(u)} = \sum_\rho T_\rho^{(v)} \frac{\partial v_\rho}{\partial u_\alpha}.$$

Hence operating with $\sum_\alpha \frac{\partial u_\alpha}{\partial v_\mu}$ * we obtain

$$T_\alpha^{(v)} = \sum_\alpha T_\alpha^{(u)} \frac{\partial u_\alpha}{\partial v_u}.$$

Thus [cf. section 8] the required tensor property is proved.

Secondly, it is evident that if the arbitrary tensor $\| S_\mu \|$ had been covariant, then $\| T^\mu \|$ would have been contravariant.

Case (ii). If $\|T_{\mu\nu}\|$ be an array of components, of the 2nd order, defined for every coordinate-system, and if, whatever 1st-order contravariant tensor $\|S^{\nu}\|$ may be, the restricted product

$$\left\| \sum_{\rho} T_{\mu\rho} S^{\rho} \right\|$$

is a covariant tensor of the 1st order, then $\|T_{\mu\nu}\|$ is a covariant tensor of the 2nd order. For by hypothesis

$$S^{\alpha}_{(u)} = \sum_{\rho} S^{\rho}_{(v)} \frac{\partial u_{\alpha}}{\partial v_{\rho}}$$

and
$$\sum_{\rho} T^{(v)}_{\mu\rho} S^{\rho}_{(v)} = \sum_{\beta} \left[\sum_{\alpha} T^{(u)}_{\beta\alpha} S^{\alpha}_{(u)} \right] \frac{\partial u_{\beta}}{\partial v_{\mu}}.$$

Hence by substituting from the former into the latter equation

$$\sum_{\rho} \left[T^{(v)}_{\mu\rho} - \sum_{\beta} \sum_{\alpha} T^{(u)}_{\beta\alpha} \frac{\partial u_{\beta}}{\partial v_{\mu}} \frac{\partial u_{\alpha}}{\partial v_{\rho}} \right] S^{\rho}_{(v)} = 0.$$

Thus, as in Case (i), by suitable choices for $\|S^{p}\|$ the tensor property is proved.

An analogous theorem holds in which invariance and contravariance are interchanged, or in which the array $\|T\|$ is proved to be a mixed tensor.

Case (iii). If $\|T_{\mu\nu}\|$ be an array of components, of the 2nd order, defined for every coordinate-system, and if, whatever 2nd order contravariant tensor $\|S^{\mu\nu}\|$ may be, the restricted product

$$\left\| \sum_{\rho} T_{\mu\rho} S^{\nu\rho} \right\|$$

is a mixed tensor of the 2nd order, then the array $\|T_{\mu\nu}\|$ is a covariant tensor of the 2nd order.

For by hypothesis

$$S^{\nu\rho}_{(v)} = \sum_{\beta} \sum_{\delta} S^{\beta\delta}_{(u)} \frac{\partial v_{\nu}}{\partial u_{\beta}} \frac{\partial v_{\rho}}{\partial u_{\delta}}.$$

Hence operating with $\sum_\rho \frac{\partial u_\gamma}{\partial v_\rho} *$, we deduce

$$\sum_\rho S_{(v)}^{\nu\rho} \frac{\partial u_\gamma}{\partial v_\rho} = \sum_\beta \sum_\delta S_{(u)}^{\beta\delta} \frac{\partial v_\nu}{\partial u_\beta} \sum_\rho \frac{\partial u_\gamma}{\partial v_\rho} \frac{\partial v_\rho}{\partial u_\delta}$$

$$= \sum_\beta S_{(u)}^{\beta\gamma} \frac{\partial v_\nu}{\partial u_\beta}.$$

Also by hypothesis

$$\sum_\rho T_{\mu\rho}^{(v)} S_{(v)}^{\nu\rho} = \sum_\alpha \sum_\beta \left[\sum_\gamma T_{\alpha\gamma}^{(u)} S_{(u)}^{\beta\gamma} \right] \frac{\partial u_\alpha}{\partial v_\mu} \frac{\partial v_\nu}{\partial u_\beta}$$

$$= \sum_\alpha \sum_\gamma T_{\alpha\gamma}^{(u)} \frac{\partial u_\alpha}{\partial v_\mu} \sum_\beta S_{(u)}^{\beta\gamma} \frac{\partial v_\nu}{\partial u_\beta}$$

$$= \sum_\rho \left[\sum_\alpha \sum_\gamma T_{\alpha\gamma}^{(u)} \frac{\partial u_\alpha}{\partial v_\mu} \frac{\partial u_\gamma}{\partial v_\rho} \right] S_{(v)}^{\nu\rho}.$$

Thus
$$\sum_\rho \left[T_{\mu\rho}^{(v)} - \sum_\alpha \sum_\gamma T_{\alpha\gamma}^{(u)} \frac{\partial u_\alpha}{\partial v_\mu} \frac{\partial u_\gamma}{\partial v_\rho} \right] S_{(v)}^{\nu\rho} = 0.$$

Hence, as above, the tensor property follows. Analogous theorems follow for suitable interchanges of the covariant and invariant types.

Case (iv). If $\|T_{\mu\nu}\|$ be an array of components, of any order, defined for every coordinate-system, and if, whatever Ist-order covariant tensor $\|S_\lambda\|$ may be, the product $\|S_\lambda T_{\mu\nu}\|$ is a covariant tensor of the 3rd order, then $\|T_{\mu\nu}\|$ is a covariant tensor of the 2nd order.

For by hypothesis

$$S_\lambda^{(v)} T_{\mu v}^{(v)} = \sum_\alpha \sum_\beta \sum_\gamma S_\alpha^{(u)} T_{\beta\gamma}^{(u)} \frac{\partial u_\alpha}{\partial v_\lambda} \frac{\partial u_\beta}{\partial v_\mu} \frac{\partial u_\gamma}{\partial v_\nu}.$$

Also by hypothesis

$$S_\lambda^{v)} = \sum_\alpha S_\alpha^{(u)} \frac{\partial u_\alpha}{\partial v_\lambda}.$$

Hence from both equations

$$\left[T_{\mu\nu}^{(v)} - \sum_{\beta}\sum_{\gamma} T_{\beta\gamma}^{(u)} \frac{\partial u_\beta}{\partial v_\mu} \frac{\partial u_\gamma}{\partial v_\nu}\right] S_\lambda^{(v)} = 0.$$

Hence, by suitable choices for the arbitrary tensor $\|S_\lambda^{(v)}\|$, the tensor property for $\|T_{\mu\nu}^{(v)}\|$ is proved.

Analogous theorems can be proved for any suitable interchanges of covariance and contravariance of type.

General Theorem. If the product, restricted (multiply or singly) or unrestricted, of an array, of any order and defined for every coordinate-system, with every arbitrary tensor of any one definite type and order be a tensor [of suitable type and order], then the array is a tensor [of suitable type and order].

It is evident that the types of proof given above for the four special cases can be adapted for every case of this general theorem.

18. **Differential Forms.** Since

$$dv_\mu = \sum_\alpha \frac{\partial v_\mu}{\partial u_\alpha} du_\alpha$$

it follows that

$$\|dv_\mu\|, \text{ and } \|dv_\mu dv_\nu\|, \text{ and } \|dv_\lambda dv_\mu dv_\nu\|, \text{ etc.,}$$

are contravariant tensors of the 1st, 2nd, 3rd, etc., orders respectively.

Hence if $\|F_\mu\|$ be a covariant tensor

$$\|\sum_\mu F_\mu^{(\mu)} du_\mu\|$$

is invariant. We adopt the notation

$$dF = \sum_\mu F_\mu^{(\mu)} du_\mu \text{ ....} (14).$$

Then dF is a differential form of the 1st order.

Similarly if $\|S_{\mu\nu}\|$ be a covariant tensor

$$\|\sum_\mu\sum_\nu S_{\mu\nu}^{(\mu)} du_\mu du_\nu\|$$

is invariant. We adopt the notation

$$dS^2 = \sum_\mu\sum_\nu S_{\mu\nu}^{(\mu)} du_\mu du_\nu \text{ ....} (15).$$

Then dS^2 is a differential form of the 2nd order.

CHAPTER XXI

TENSORS OF THE SECOND ORDER

IT is proposed in this chapter to bring together some of the simpler notations and theorems relating to tensors of the second order.

19. **Symmetric Tensors.** The covariant and contravariant tensors $\|S_{\mu\nu}\|$ and $\|T^{\mu\nu}\|$ are respectively called 'symmetric' if in every measure-system

and
$$\left.\begin{aligned} S_{\mu\nu} &= S_{\nu\mu} \\ T^{\mu\nu} &= T^{\nu\mu} \end{aligned}\right\} \left[\mu, \nu = 1, 2, 3, 4\right] \quad \text{.......................(16).}$$

If a tensor is symmetric in one measure-system, it is symmetric in every measure-system.

For in measure-system 'u' let

$$S_{\mu\nu}^{(u)} = S_{\nu\mu}^{(u)}, \ \left[\mu, \nu = 1, 2, 3, 4\right].$$

Then

$$\begin{aligned} S_{\mu\nu}^{(v)} &= \sum_{\alpha}\sum_{\beta} S_{\alpha\beta}^{(u)} \frac{\partial u_\alpha}{\partial v_\mu} \frac{\partial u_\beta}{\partial v_\nu} \\ &= \sum_{\alpha}\sum_{\beta} S_{\beta\alpha}^{(u)} \frac{\partial u_\beta}{\partial v_\nu} \frac{\partial u_\alpha}{\partial v_\mu} \\ &= S_{\nu\mu}^{(v)}. \end{aligned}$$

The theorem holds for contravariant tensors with suitable interchanges of the covariant and contravariant types.

We notice that in the case of the differential form dS^2 in the preceding section there is no loss of generality in considering the tensor $\|S_{\mu\nu}\|$ to be symmetrical.

20. **Skew Tensors.** The covariant and contravariant tensors $\|S_{\mu\nu}\|$ and $\|T^{\mu\nu}\|$ are respectively called 'skew' if in every measure-system

$$S_{\mu\nu} + S_{\nu\mu} = 0, \text{ and } + T^{\mu\nu} + T^{\nu\mu} = 0 \text{..................}(17).$$

If a tensor is skew in one measure-system, it is skew in every measure-system.

For in measure-system 'u' let

$$S^{(u)}_{\mu\nu} + S^{(u)}_{\nu\mu} = 0.$$

Then

$$S^{(v)}_{\mu\nu} + S^{(v)}_{\nu\mu} = \sum_{\alpha}\sum_{\beta} S^{(u)}_{\alpha\beta} \frac{\partial u_\alpha}{\partial v_\mu} \frac{\partial u_\beta}{\partial v_\nu} + \sum_{\beta}\sum_{\alpha} S^{(u)}_{\beta\alpha} \frac{\partial u_\beta}{\partial v_\nu} \frac{\partial u_\alpha}{\partial v_\mu}$$

$$= \sum_{\alpha}\sum_{\beta} \left(S^{(u)}_{\alpha\beta} + S^{(u)}_{\beta\alpha}\right) \frac{\partial u_\alpha}{\partial v_\mu} \frac{\partial u_\beta}{\partial v_\nu}$$

$$= 0.$$

An analogous proof holds for $\|T^{\mu\nu}.\|$

It is evident that for skew tensors

$$S_{\mu\mu} = 0, \quad T^{\mu\mu} = 0 \text{..........................}(17 \cdot 1).$$

21. **The Determinants.** If $\|S_{\mu\nu}\|$ and $\|T^{\mu\nu}\|$ be respectively covariant and contravariant tensors of the second order, the symbols $S^{(u)}$ and $T_{(u)}$ represent the determinants formed by the components as elements, so that

$$S^{(u)} = \text{det.}\,|S^{(u)}_{\mu\nu}| \text{................................}(18),$$

and

$$T_{(u)} = \text{det.}\,|T^{\mu\nu}_{(u)}| \text{............................}(18.1).$$

It at once follows from the law of the multiplication of determinants that

$$S^{(v)} = S^{(u)} \times \left\{\frac{\partial(u_1, u_2, u_3, u_4)}{\partial(v_1, v_2, v_3, v_4)}\right\}^2 \text{....................}(19),$$

and $$T_{(v)} = T_{(u)} \times \left\{ \frac{\partial\left(v_1, v_2, v_3, v_4\right)}{\partial\left(u_1, u_2, u_3, u_4\right)} \right\}^2 \quad \text{.................}(19{\cdot}1).$$

A tensor is called 'special' if its determinant vanishes. It is evident from the above equation that if a tensor be special in one coordinate-system, it is special in every coordinate-system.

Since we are considering a four-dimensional manifold, a skew tensor is not necessarily special. But in a three-dimensional manifold every skew tensor would be special.

If $\|M^{\nu}{}_{\mu}\|$ be a mixed tensor, the symbol M will denote the determinant formed by the components as elements, so that

$$M = \text{det.} \mid M^{\nu}{}_{\mu}(x) \mid \text{..........................}(19{\cdot}2).$$

It is unnecessary to denote the coordinate-system 'x' in the symbol for the determinant since the value of the determinant is the same in all coordinate-systems, that is to say, the determinant is an invariant. For if $M7$ be the value in system 'y' and M in system 'x,'

$$\begin{aligned} M' &= M \times \frac{\partial\left(y_1, y_2, y_3, y_4\right)}{\partial\left(x_1, x_2, x_3, x_4\right)} \times \frac{\partial\left(x_1, x_2, x_3, x_4\right)}{\partial\left(y_1, y_2, y_3, y_4\right)} \\ &= M \quad \text{...}(19.3). \end{aligned}$$

22. **Associate Tensors.** Let $\|S_{\mu\nu}\|$ and $\|T^{\mu\nu}\|$ be a pair of tensors of the second order, one covariant and the other contravariant, such that in the coordinate-system 'u'

$$\sum_{\rho} T^{\mu\rho}_{(u)} S^{(u)}_{\nu\rho} = I^{\mu}_{\nu} \text{................................}(20),$$

then the analogous property holds for every coordinate-system.

For

$$\begin{aligned} \sum_{\rho} T^{\mu\rho}_{(v)} S^{(v)}_{\nu\rho} &= \sum_{\alpha}\sum_{\beta}\sum_{\gamma}\sum_{\delta} T^{\alpha\beta}_{(u)} S^{(u)}_{\gamma\delta} \frac{\partial v_\mu}{\partial u_\alpha}\frac{\partial u_\gamma}{\partial v_\nu} \sum_{\rho} \frac{\partial v_\rho}{\partial u_\beta}\frac{\partial u_\delta}{\partial v_\rho} \\ &= \sum_{\alpha}\sum_{\gamma} \left[\sum_{\beta} T^{\alpha\beta}_{(u)} S^{(u)}_{\gamma\beta} \right] \frac{\partial v_\mu}{\partial u_\alpha}\frac{\partial u_\gamma}{\partial v_\nu} \end{aligned}$$

$$= \sum_\alpha \sum_\gamma I_\gamma^\alpha \frac{\partial v_\mu}{\partial u_\alpha} \frac{\partial u_\gamma}{\partial v_\nu}$$

$$= \sum_\alpha \frac{\partial v_\mu}{\partial u_\alpha} \frac{\partial u_\alpha}{\partial v_\nu}$$

$$= I_\nu^\mu.$$

A pair of tensors with this property are called 'Associate Tensors.'

If either of the two tensors be not special, it has one and only one associate tensor which is also not special.

In the sequel, unless it is otherwise expressly stated, in dealing with associate tensors we shall always assume that we are considering non-special tensors.

Thus the associate of the tensor associated with a given tensor is the original given tensor. The associate of any tensor $\|S_{\mu\nu}\|$ will be denoted by $\|S^{\mu\nu}\|$, and conversely. Also with the above notation,

$$S_{(u)}^{\mu\nu} = \left[\text{cofactor of } S_{\mu\nu}^{(u)} \text{ in } S^{(u)}\right] \div S^{(u)} \quad \text{..............(20.1)},$$

and $$S_{\mu\nu}^{(u)} = \left[\text{cofactor of } S_{(u)}^{\mu\nu} \text{ in } S_{(u)}\right] \div S_{(u)} \quad \text{..............(20.2)},$$

and $S_{(u)} S^{(u)} = 1$

Associate tensors enable us to solve tensor equations of the form

$$\sum_\rho S_{\mu\rho}^{(u)} X_{\cdots}^{\cdots\rho} = D_{\mu\cdots}^{\cdots}, \left[\mu = 1, 2, 3, 4\right] \text{......................(21)}.$$

For operating with $\sum_\mu S_{(u)}^{\mu\nu} *$, we find

$$\sum_\rho I_\rho^\nu X_{\cdots}^{\cdots\rho} = \sum_\mu S_{(u)}^{\mu\nu} D_{\mu\cdots}^{\cdots},$$

i.e. $$X_{\cdots}^{\cdots\nu} = \sum_\mu S_{(u)}^{\mu\nu} D_{\mu\cdots}^{\cdots}.$$

Analogously we can solve

$$\sum_\rho S_{(u)}^{\mu\rho} X_{\cdots\rho}^{\cdots} = D_{\cdots}^{\mu\cdots}, \left[\mu = 1, 2, 3, 4\right] \text{.................(21·1)}.$$

The theory of associate tensors applies also to mixed tensors. For, exactly as above, if $\|S_\mu^\nu\|$ and $\|T_\mu^\nu\|$ be a pair of mixed tensors of the second order, such that in one coordinate-system

$$\sum_\rho S^\rho_\mu T^\nu_\rho = I^\nu_\mu \quad \text{..................................(22)},$$

then the property holds for every coordinate-system. Also all the analogous theorems hold.

The associate of a non-special mixed tensor $\|S^\mu_\nu\|$ will be written $\|\bar{S}^\mu_\nu\|$

Thus $$\|\bar{\bar{S}}^\mu_\nu\| = \|S^\mu_\nu\| \quad \text{..................................(22.1)}$$

Either both or neither of a pair of associate tensors (invariant or covariant) are symmetrical, and either both or neither are skew.

For if $\|S_{\mu\nu}\|$ be symmetrical, the cofactor of $S^{(u)}_{\mu\nu}$ in $S^{(u)}$ is equal to the cofactor of $S^{(u)}_{\mu\nu}$ in $S^{(u)}$, and analogously for contravariance.

Also if $\|S_{\mu\nu}\|$ be skew, the cofactor of $S^{(u)}_{\mu\nu}$ in $S^{(u)}$ is equal to the negative of the cofactor of $S^{(u)}_{\mu\nu}$ in $S^{(u)}$ and analogously for contravariance.

23. **Derivative Tensors.** By means of a pair of tensors of the second order we can derive tensors of various types which are called tensors derivative from the given tensor.

For example, let $\|S_{\mu\nu}\|$ and $\|S^{\mu\nu}\|$ be one pair of associate tensors and $\|T_{\mu\nu}\|$ and $\|T^{\mu\nu}\|$ be another pair. Then the derivative tensors

$$\left.\begin{array}{l} \|\sum_\rho S_{\mu\rho} T^{\nu\rho}\| \text{ and } \|\sum_\rho S_{\rho\mu} T^{\rho\nu}\| \\ \text{and} \quad \|\sum_\rho S_{\mu\rho} T^{\rho\nu}\| \text{ and } \|\sum_\rho S_{\rho\mu} T^{\nu\rho}\| \end{array}\right\} \quad \text{......................(23)}$$

are derivative tensors of mixed type. If one of the tensors $\|S_{\mu\nu}\|$ or $\|T_{\mu\nu}\|$ be symmetrical these four tensors coalesce into two tensors, and if both be symmetrical they coalesce into one tensor.

Similarly $$\|\sum_\rho \sum_\sigma T^{\mu\rho} T^{\nu\sigma} S_{\rho\sigma}\| \quad \text{..............................(23·1)}$$

is a contravariant derivative tensor. There are four derivative tensors of this sort which coalesce into three if $\|S_{\mu\nu}\|$ be symmetrical, and into one if $\|T_{\mu\nu}\|$ be symmetrical.

Again $$\|\sum_\rho \sum_\sigma T_{\mu\rho} T_{\nu\sigma} S^{\rho\sigma}\| \quad \text{............................(23·2)}$$

is a covariant derivative tensor, with other analogous tensors of the same sort.

Finally there are analogous sets of derivative tensors in which $\|S_{\mu\nu}\|$ and $\|T_{\mu\nu}\|$ have interchanged rôles.

CHAPTER XXII

THE GALILEAN TENSORS

24. **Galilean Tensors.** We have already [cf. Chapter IV of Part I] defined the symbol

$$\omega_\mu^2, \quad [\mu = 1, 2, 3, 4]$$

by

$$\left.\begin{aligned} &\omega_\mu^2, = 1, \quad [\mu = 1, 2, 3] \\ &\omega_4^2 = -c^2 \end{aligned}\right\} \qquad \text{.............................(24).}$$

Define the Galilean tensor [i.e. $\|G\{u\}_{\mu\nu}\|$] of the coordinate-system 'u' by

$$\left.\begin{aligned} &G\,\{u\}_{\mu\nu}^{(u)} = 0, \quad [\mu \neq \nu] \\ &G\,\{u\}_{\mu\mu}^{(u)} = -\omega_\mu^2, \quad [\mu = 1, 2, 3, 4] \end{aligned}\right\} \qquad \text{................(24·1).}$$

Then in any other coordinate-system 'v,'

$$\begin{aligned} G\,\{u\}_{\mu\nu}^{(v)} &= \sum_\alpha \sum_\beta G\,\{u\}_{\alpha\beta}^{(u)} \frac{\partial u_\alpha}{\partial v_\mu} \frac{\partial u_\beta}{\partial v_\nu} \\ &= -\sum_\alpha \omega_\alpha^2 \frac{\partial u_\alpha}{\partial v_\mu} \frac{\partial u_\alpha}{\partial v_\nu} \qquad \text{..................(24·2)} \end{aligned}$$

We will always assume that in any coordinate-system 'v' the coordinate-type which is to play the part of the exceptional axis for the Galilean tensor is to be assigned the subscript 4.

With this convention, the condition that the coordinate-systems 'u' and 'v' have the same Galilean tensor is

$$\left.\begin{aligned}\sum_\alpha \omega_\alpha^2 \frac{\partial u_\alpha}{\partial v_\mu}\frac{\partial u_\alpha}{\partial v_\nu} &= 0, \quad [\mu \neq \nu] \\ \sum_\alpha \omega_\alpha^2 \frac{\partial u_\alpha}{\partial v_\mu}\frac{\partial u_\alpha}{\partial v_\mu} &= \omega_\mu^2\end{aligned}\right\} \quad \ldots\ldots\ldots\ldots(25).$$

Operate on $\sum_\alpha \omega_\alpha^2 \frac{\partial u_\alpha}{\partial v_\mu}\frac{\partial u_\alpha}{\partial v_\nu}$ with $\sum_\nu \frac{\partial v_\nu}{\partial u_\beta}$ *.

Then, from the two equations above, we obtain

$$\sum_\alpha \omega_\alpha^2 \frac{\partial u_\alpha}{\partial v_\mu}\sum_\nu \frac{\partial u_\alpha}{\partial v_\mu}\frac{\partial v_\nu}{\partial u_\beta} = \omega_\mu^2 \frac{\partial v_\mu}{\partial u_\beta},$$

i.e.

$$\left.\omega_\beta^2 \frac{\partial u_\beta}{\partial v_\mu} = \omega_\mu^2 \frac{\partial v_\mu}{\partial u_\beta}, \quad [\mu, \beta = 1, 2, 3, 4]\right\} \quad \ldots\ldots\ldots(25{\cdot}1).$$

25. **Galilean Differential Forms.** The differential form arising from this Galilean tensor is

$$\left.\begin{aligned}dG^2 &= \sum_\mu\sum_\nu G\{u\}_{\mu\nu}^{(p)} dp_\mu dp_\nu \\ &= -\sum_\alpha \omega_\alpha^2 du_\alpha^2\end{aligned}\right\} \quad \ldots\ldots\ldots\ldots(26).$$

It must be remembered that this particular Galilean differential form has the Galilean property for the group of coordinate-systems, such as 'v,' which are connected with the coordinate-system 'u' by sets of equations of the type of equations (25·1) above. Call such a group of coordinate-systems a 'Galilean group.' It is evident that a Galilean group is defined by any one of the coordinate-systems which belong to it, since each such system belongs to one and only one such group.

26. **The Linear Equations of Transformation.** Let a track in the manifold be defined by considering (u_1, u_2, u_3) as appropriate functions of u_4, and with this supposition let $(\dot{u}_1, \dot{u}_2, \dot{u}_3)$ stand for $\left(\frac{du_1}{du_4}, \frac{du_2}{du_4}, \frac{du_3}{du_4}\right)$.

We now seek the condition that

$$\int_A^B \surd dG^2,$$

along a track between any given pair of event-particles A and B of the manifold, may have a stationary value. This is given by the adaptation to this case of equations (7) of Part II Chapter v. Since the coordinate-system 'u' is a member of the relevant Galilean group, these equations reduce to

$$\frac{d}{du_4}\frac{\dot{u}_\mu}{\sqrt{\left\{1-\dfrac{\dot{u}_1^2+\dot{u}_2^2+\dot{u}_3^2}{c^2}\right\}}}=0, \quad [\mu=1,2,3]$$

i.e. $$u_\mu = a_\mu u_4 + b_\mu, \quad [\mu=1,2,3] \quad \text{...............(27)},$$

where α_μ and b_μ are constants.

If the coordinate-system 'v' be another member of the same Galilean group, the same track, from A and B and stationary, must be expressible in the form

$$v_\rho = c_\rho v_4 + d_\rho, \quad [\rho=1,2,3] \quad \text{......................(7·1)}.$$

where c_ρ and d_ρ are constants.

Hence the equations of transformation relating any pair of coordinate-systems 'u' and 'v' belonging to the same Galilean group must be of the linear form

$$\omega_\mu(v_\mu - e_\mu) = \sum_\alpha l_{\mu\alpha}\omega_\alpha u_\alpha \text{.........................(28)},$$

where e_μ and $l_{\mu a}$ [μ, a = 1, 2, 3, 4] are constants.

Furthermore, from equations (25) above [interchanging 'u' and 'v' in their application],

$$\left.\begin{aligned}\sum_\mu l_{\mu\alpha}l_{\mu\beta} &= 0, \quad [\alpha \neq \beta] \\ &= 1, \quad [\alpha = \beta]\end{aligned}\right\}\text{........................(28·1)}.$$

Also we can conceive equations (28) to be solved in the form

$$\omega_\alpha(u_\alpha - f_\alpha) = \sum_\mu l'_{\alpha\mu}\omega_\mu v_\mu \text{ ....(28·2)}.$$

Hence
$$\omega_\beta^2 \frac{\partial u_\beta}{\partial v_\mu} = \omega_\beta\, \omega_\mu l'_{\beta\mu},$$

and
$$\omega_\mu^2 \frac{\partial v_\mu}{\partial u_\beta} = \omega_\mu\, \omega_\beta l_{\mu\beta}.$$

Thus [cf. equations (25.1)]

$$l'_{\beta\mu} = l'_{\mu\beta}.$$

27. **Cartesian Group.** Thus the Cartesian coordinate-systems of Chapter IV of Part I are a particular Galilean group of coordinate-systems [such as the system '*x*'] which have a peculiar spatio-temporal significance in the four-dimensional continuum of nature. When we are discussing the Galilean tensor of this group, we symbolise it by $\|G_{\mu\nu}\|$ in place of the longer $\|G\{x\}_{\mu\nu}\|$.

We will call this Galilean group of coordinate-systems the 'Cartesian group,' and the corresponding Galilean tensor is (in case of doubt) called the 'Cartesian Galilean tensor.' Furthermore, in discussing Galilean tensors we will habitually consider in illustration the Cartesian group and its Galilean tensor. But the theorems are quite general and hold for any Galilean group.

28. **Associate Galilean Tensors and Galilean Derivative Tensors.** Let the coordinate-system '*x*' belong to the Cartesian group. Then the associate of the Galilean tensor is $\|G^{\mu\nu}\|$ where

$$\left.\begin{aligned} G^{\mu\nu}_{(x)} &= 0, \quad [\mu \neq \nu] \\ G^{\mu\mu}_{(x)} &= -\frac{1}{\omega_\mu^2}, \quad [\mu = 1, 2, 3, 4] \end{aligned}\right\} \text{......................(29)}.$$

By means of the Galilean tensor and its associate tensor derivative tensors are found from any given pair of associate tensors, $\|S_{\mu\nu}\|$ and $\|S^{\mu\nu}\|$, which are called the 'Galilean derivatives' from $\|S_{\mu\nu}\|$ or $\|S^{\mu\nu}\|$.

Thus the Galilean derivatives of mixed type are

$$\|\sum_\rho G^{\nu\rho} S_{\mu\rho}\| \text{ and } \|\sum_\rho G_{\mu\rho} S^{\nu\rho}\| \text{......................(30)}.$$

The components in the coordinate-system 'x' of the former tensor [μ the covariant affix] are

$$G^{\nu\nu}_{(x)} S^{(x)}_{\mu\nu}, \quad \left[\mu, \nu = 1, 2, 3, 4\right]$$

i.e.
$$\left.\begin{array}{ll} -S^{(x)}_{\mu\nu} & \left[\mu = 1, 2, 3, 4; \nu = 1, 2, 3\right] \\ \text{and} \quad \frac{1}{c^2} S^{(x)}_{\mu 4}, & \left[\mu = 1, 2, 3, 4\right] \end{array}\right\} \quad \ldots\ldots\ldots\ldots\ldots (30 \cdot 1).$$

The components in the coordinate-system 'x' of the latter tensor [μ the covariant affix] are

$$G^{(x)}_{\mu\mu} S^{\nu\mu}_{(x)}, \quad \left[\mu, \nu = 1, 2, 3, 4\right],$$

i.e.
$$\left.\begin{array}{ll} -S^{\nu\mu}_{(x)}, & \left[\nu = 1, 2, 3, 4; \mu = 1, 2, 3\right] \\ \text{and} \quad c^2 S^{\nu 4}_{(x)}, & \left[\nu = 1, 2, 3, 4\right] \end{array}\right\} \quad \ldots\ldots\ldots\ldots\ldots (30.2)$$

The Galilean derivative of contravariant type is

$$\left\| \sum_\rho \sum_\sigma G^{\mu\rho} G^{\nu\sigma} S_{\rho\sigma} \right\| \quad \ldots\ldots\ldots\ldots\ldots\ldots\ldots\ldots (31).$$

The components in the coordinate-system 'x' of this tensor are

$$G^{\mu\mu}_{(x)} G^{\nu\nu}_{(x)} S^{(x)}_{\mu\nu}, \quad \left[\mu, \nu = 1, 2, 3, 4\right],$$

i.e.
$$S^{(x)}_{\mu\nu}, \quad \left[\mu, \nu = 1, 2, 3\right] \quad \ldots\ldots\ldots\ldots\ldots\ldots (31 \cdot 1).$$

and
$$-\frac{1}{c^2} S^{(x)}_{\mu 4}, \quad \left[\mu = 1, 2, 3; \nu = 4\right] \quad \ldots\ldots\ldots (31 \cdot 2).$$

and
$$-\frac{1}{c^2} S^{(x)}_{4\nu}, \quad \left[\mu = 4; \nu = 1, 2, 3\right] \quad \ldots\ldots\ldots (31 \cdot 3).$$

and
$$\frac{1}{c^4} S^{(x)}_{44}, \quad \left[\mu = 4; \nu = 4\right] \quad \ldots\ldots\ldots\ldots\ldots (31 \cdot 4).$$

The components of this contravariant Galilean derivative are linear functions of the components of $S\mu\nu$.

The Galilean derivative of covariant type is

$$\left\| \sum_\rho \sum_\sigma G_{\mu\rho} G_{\nu\sigma} S^{\rho\sigma} \right\| \quad \ldots\ldots\ldots\ldots\ldots\ldots\ldots\ldots (32).$$

The components in the coordinate-system 'x' of this tensor are

$$G_{\mu\mu}^{(x)} G_{\nu\nu}^{(x)} S_{(x)}^{\mu\nu}, \quad [\mu, \nu = 1, 2, 3, 4],$$

i.e.

$$S_{(x)}^{\mu\nu}, \quad [\mu, \nu = 1, 2, 3] \quad \text{.................... (32·1).}$$

and

$$-c^2 S_{(x)}^{\mu 4}, \quad [\mu = 1, 2, 3; \nu = 4] \quad \text{.............. (32·2).}$$

and

$$-c^2 S_{(x)}^{4\nu}, \quad [\mu = 4; \nu = 1, 2, 3] \quad \text{.............. (32·3).}$$

and

$$c^4 S_{(x)}^{44}, \quad [\mu = 4; \nu = 4] \quad \text{.................... (32·4).}$$

The components of this covariant Galilean derivative are linear functions of the components of $\|S^{\mu\nu}\|$.

Finally the Galilean invariants are

$$\sum_\rho \sum_\sigma G^{\rho\sigma} S_{\rho\sigma} \text{ and } \sum_\rho \sum_\sigma G_{\rho\sigma} S^{\rho\sigma} \quad \text{...................... (33).}$$

Thus in any coordinate-system 'x' of the Cartesian group there are the two group invariants

$$S_{11}^{(x)} + S_{22}^{(x)} + S_{33}^{(x)} - \frac{1}{c^2} S_{44}^{(x)} \quad \text{...................... (33.1)}$$

and

$$S_{(x)}^{11} + S_{(x)}^{22} + S_{(x)}^{33} - c^2 S_{(x)}^{44} \quad \text{...................... (33.2).}$$

29. **Galilean Derivative Tensors of the First Order.** Let $\|F_\mu\|$ be a covariant tensor of the first order, then its 'Galilean derivative' is the contravariant tensor

$$\|\sum_\rho G^{\mu\rho} F_\rho\| \quad \text{.................................... (34).}$$

The components in the Cartesian coordinate-system 'x' of this tensor are

$$G_{(x)}^{\mu\mu} F_\mu^{(x)}, \quad [\mu = 1, 2, 3, 4] \quad \text{............... (34·1).}$$

i.e.

$$-F_\mu^{(x)}, \quad [\mu = 1, 2, 3] \quad \text{.................. (34·2).}$$

and

$$\frac{1}{c^2} F_4^{(x)}, \quad [\mu = 4] \quad \text{...................... (34·3).}$$

The Galilean invariant is

$$\sum_\rho \sum_\sigma G^{\rho\sigma} F_\rho F_\sigma \quad \text{.............................. (34·4).}$$

Thus in any coordinate-system 'x' of the Cartesian group there is the group invariant

$$\{F_1^{(x)}\}^2 + \{F_2^{(x)}\}^2 + \{F_3^{(x)}\}^2 - \frac{1}{c^2}\{F_4^{(x)}\}^2 \text{ ....(34·5).}$$

Again let $\|F^\mu\|$ be a contravariant tensor of the first order, then its 'Galilean derivative' is the covariant tensor

$$\|\sum_\rho G_{\mu\rho} F^\rho\| \text{ ....(35).}$$

The components in the Cartesian coordinate-system 'x' of this tensor are

$$G_{\mu\mu}^{(x)} F_{(x)}^{\mu}, \left[\mu = 1, 2, 3, 4\right] \text{ ....(35·1).}$$

i.e.

$$-F_{(x)}^{\mu}, \left[\mu = 1, 2, 3\right] \text{ ....(35·2).}$$

and

$$c^2 F_{(x)}^{4}, \left[\mu = 4\right] \text{ ....(35·3).}$$

The Galilean invariant is

$$\sum_\rho \sum_\sigma G_{\rho\sigma} F^\rho F^\sigma \text{ ....(36).}$$

Thus in any coordinate-system 'x' of the Cartesian group there is the group invariant

$$\left\{F_{(x)}^{1}\right\}^2 + \left\{F_{(x)}^{2}\right\}^2 + \left\{F_{(x)}^{3}\right\}^2 - c^2\left\{F_{(x)}^{4}\right\}^2 \text{ ....(36·1)}$$

CHAPTER XXIII

THE DIFFERENTIATION OF TENSOR COMPONENTS

30. **The Christoffel Three-Index Symbols.** Let $\|H_{\mu\nu}\|$ be any symmetric covariant tensor. The Christofiel Three-Index Symbol of the first kind is defined by

$$\left. \begin{array}{c} H\left[\mu\nu,\lambda\right]^{(u)} = \dfrac{1}{2}\left\{\dfrac{\partial H_{\mu\lambda}^{(u)}}{\partial u_\nu} + \dfrac{\partial H_{\nu\lambda}^{(u)}}{\partial u_\mu} - \dfrac{\partial H_{\mu\nu}^{(u)}}{\partial u_\lambda}\right\} \\ \left[\lambda,\mu,\nu = 1,2,3,4\right] \end{array} \right\} \text{..........(37).}$$

The Christoffel Three-Index Symbol of the second kind is defined by

$$H\left\{\mu\nu,\lambda\right\}^{(u)} = \sum_\rho H^{\lambda\rho} H\left[\mu\nu,\rho\right]^{(u)} \quad \text{................(37·1)}$$

Then

$$\begin{aligned} \sum_\alpha H_{\sigma\lambda}^{(u)} H\left\{\mu\nu,\sigma\right\}^{(u)} &= \sum_\rho\sum_\sigma H_{\sigma\lambda}^{(u)} H^{\sigma\rho} H\left[\mu\nu,\rho\right]^{(u)} \\ &= H\left[\mu\nu,\lambda\right]^{(u)} \quad \text{....................(37·2).} \end{aligned}$$

It is evident that

$$H\left[\mu\nu,\lambda\right]^{(u)} = H\left[\nu\mu,\lambda\right]^{(u)} \text{..........................(38)}$$

and

$$H\left\{\mu\nu,\lambda\right\}^{(u)} = H\left\{\nu\mu,\lambda\right\}^{(u)} \quad \text{......................(38.1)}$$

In general neither $H[\mu v, \lambda]^{(u)}$ nor $H\{\mu v, \lambda\}^{(u)}$ is a tensor, though we shall prove that they are group tensors for any Galilean group.

31. **Differentiation of Determinants of Tensors.** Let the covariant tensor $\|S_{\mu\nu}\|$ be infinitesimally increased to $\|S_{\mu\nu} + \delta S_{\mu\nu}\|$, and in consequence let $S^{(u)}$ become $\delta S^{(u)}$. Then

$$\delta S^{(u)} = \sum_{\mu}\sum_{\nu} \left\{\text{confactor of } S_{\mu\nu}^{(u)}\right\} \times \delta S_{\mu\nu}^{(u)}$$

$$= S^{(u)} \sum_{\mu}\sum_{\nu} S_{(u)}^{\mu\nu} \times \delta S_{\mu\nu}^{(u)} \quad \text{.........................} (39).$$

Thus $$\frac{\partial S^{(u)}}{\partial u_\lambda} = S^{(u)} \sum_{\mu}\sum_{\nu} S_{(u)}^{\mu\nu} \frac{\partial S_{\mu\nu}^{(u)}}{\partial u_\lambda} \quad \text{........................} (39.1).$$

Analogously, if the contravariant tensor $\|T_{(\mu)}^{\mu\nu}\|$ be infinitesimally increased to $\|T_{(\mu)}^{\mu\nu}\| + \|\delta T_{(\mu)}^{\mu\nu}\|$, and in consequence $T_{(u)}$ increases to $T_{(u)} + \delta T_{(u)}$, then

$$\delta T_{(u)} = T_{(u)} \sum_{\mu}\sum_{\nu} T_{\mu\nu}^{(u)} \times \delta T_{(u)}^{\mu\nu} \quad \text{................} (40),$$

and $$\frac{\partial T_{(u)}}{\partial u_\lambda} = T_{(u)} \sum_{\mu}\sum_{\nu} T_{\mu\nu}^{(u)} \frac{\partial T_{(u)}^{\mu\nu}}{\partial u_\lambda} \quad \text{.............} (41.1).$$

Now recur to the symmetric co variant tensor $\|H_{\mu\nu}\|$. Then

$$\sum_{\rho} H\{\mu\rho, \rho\}^{(u)} = \sum_{\rho}\sum_{\sigma} H_{(u)}^{\rho\sigma} H\left[\mu\rho, \sigma\right]^{(u)}$$

$$= \frac{1}{2}\sum_{\rho}\sum_{\sigma} H_{(u)}^{\rho\sigma}\left[\frac{\partial H_{\mu\sigma}^{(u)}}{\partial u_\rho} + \frac{\partial H_{\rho\sigma}^{(u)}}{\partial u_\mu} - \frac{\partial H_{\mu\rho}^{(u)}}{\partial u_\sigma}\right].$$

Now the interchange of the symbols p and σ does not affect the value of

$$\sum_{\rho}\sum_{\sigma} H_{(u)}^{\rho\sigma} \frac{\partial H_{\mu\sigma}^{(u)}}{\partial u_\rho}.$$

Hence [cf. equation (39·1)]

$$\sum_{\rho} H\left\{\mu\rho, \rho\right\}^{(u)} = \frac{1}{2}\sum_{\rho}\sum_{\sigma} H_{(u)}^{\rho\sigma} \frac{\partial H_{\rho\sigma}^{(u)}}{\partial u_\mu}$$

$$= \frac{\partial}{\partial u_\mu}\log\left\{-H^{(u)}\right\}^{\frac{1}{2}} \quad \text{..................(41).}$$

32. **The Standard Formulae.** There are certain standard formulae which are the foundation of the theory of the differentiation of tensor components.

We consider the symmetric covariant tensor $\|H_{\mu\nu}\|$.

Now
$$H_{\mu\nu}^{(v)} = \sum_{\beta}\sum_{\gamma} H_{\beta\gamma}^{(u)} \frac{\partial u_\beta}{\partial v_\mu}\frac{\partial u_\gamma}{\partial v_\nu},$$

and
$$\frac{\partial}{\partial v_\lambda} = \sum_{\alpha} \frac{\partial u_\alpha}{\partial v_\lambda}\frac{\partial}{\partial u_\alpha}.$$

Then, remembering that $\|H_{\mu\nu}\|$ is symmetric,

$$\frac{\partial H_{\mu\nu}^{(v)}}{\partial v_\lambda} = \sum_{\alpha}\sum_{\beta}\sum_{\gamma} \frac{\partial H_{\beta\gamma}^{(u)}}{\partial u_\alpha}\frac{\partial u_\alpha}{\partial v_\lambda}\frac{\partial u_\beta}{\partial v_\mu}\frac{\partial u_\gamma}{\partial v_\nu}$$
$$+\sum_{\beta}\sum_{\gamma} H_{\beta\gamma}^{(u)}\left\{\frac{\partial^2 u_\beta}{\partial v_\lambda \partial v_\mu}\frac{\partial u_\gamma}{\partial v_\nu} + \frac{\partial^2 u_\beta}{\partial v_\lambda \partial v_\nu}\frac{\partial u_\gamma}{\partial v_\mu}\right\}.$$

Hence interchanging λ, μ, v cyclically, and in the former of the two summations also interchanging α, β, γ cyclically, but in the latter retaining β and γ in their original functions, we find

$$\frac{\partial H_{\nu\lambda}^{(v)}}{\partial v_\mu} = \sum_{\alpha}\sum_{\beta}\sum_{\gamma} \frac{\partial H_{\gamma\alpha}^{(u)}}{\partial u_\beta}\frac{\partial u_\beta}{\partial v_\mu}\frac{\partial u_\gamma}{\partial v_\nu}\frac{\partial u_\alpha}{\partial v_\lambda}$$
$$+\sum_{\beta}\sum_{\gamma} H_{\beta\gamma}^{(u)}\left\{\frac{\partial^2 u_\beta}{\partial v_\mu \partial v_\nu}\frac{\partial u_\gamma}{\partial v_\lambda} + \frac{\partial^2 u_\beta}{\partial v_\mu \partial v_\lambda}\frac{\partial u_\gamma}{\partial v_\nu}\right\},$$

and
$$\frac{\partial H_{\lambda\mu}^{(v)}}{\partial v_\nu} = \sum_{\alpha}\sum_{\beta}\sum_{\gamma} \frac{\partial H_{\alpha\beta}^{(u)}}{\partial u_\gamma}\frac{\partial u_\gamma}{\partial v_\nu}\frac{\partial u_\alpha}{\partial v_\lambda}\frac{\partial u_\beta}{\partial v_\mu}$$

$$+\sum_\beta \sum_\gamma H_{\beta\gamma}^{(u)} \left\{ \frac{\partial^2 u_\beta}{\partial v_\nu \partial v_\lambda} \frac{\partial u_\gamma}{\partial v_\mu} + \frac{\partial^2 u_\beta}{\partial v_\nu \partial v_\mu} \frac{\partial u_\gamma}{\partial v_\lambda} \right\}.$$

Hence by combining these three equations

$$H\left[\mu\nu, \lambda\right]^{(v)} = \sum_\alpha \sum_\beta \sum_\gamma H\left[\beta\gamma, \alpha\right]^{(u)} \frac{\partial u_\alpha}{\partial v_\lambda} \frac{\partial u_\beta}{\partial v_\mu} \frac{\partial u_\gamma}{\partial v_\nu}$$

$$+\sum_\beta \sum_\gamma H_{\beta\gamma}^{(u)} \frac{\partial^2 u_\beta}{\partial v_\mu \partial v_\nu} \frac{\partial u_\gamma}{\partial v_\lambda} \quad \ldots\ldots\ldots\ldots\ldots\ldots (42).$$

This formula relates the three-index symbols of the first type, as expressed in different coordinate-systems.

Operate on this formula with

$$\sum_\lambda \sum_\rho H_{(v)}^{\lambda\rho} \frac{\partial u_\epsilon}{\partial v_\rho} * .$$

We consider separately the effect of this operation on each of the three terms of the above formula,

$$\sum_\lambda \sum_\rho H_{(v)}^{\lambda\rho} \frac{\partial u_\epsilon}{\partial v_\rho} H\left[\mu\nu, \lambda\right]^{(v)} = \sum_\rho \frac{\partial u_\epsilon}{\partial v_\rho} H\left\{\mu\nu, \lambda\right\}^{(v)}.$$

And

$$\sum_\alpha \sum_\beta \sum_\gamma H\left[\beta\gamma, \alpha\right]^{(u)} \frac{\partial u_\beta}{\partial v_\mu} \frac{\partial u_\gamma}{\partial v_\nu} \sum_\lambda \sum_\rho H_{(v)}^{\lambda\rho} \frac{\partial u_\epsilon}{\partial v_\rho} \frac{\partial u_\alpha}{\partial v_\lambda}$$

$$= \sum_\beta \sum_\gamma \left[\sum_\alpha H_{(u)}^{\epsilon\alpha} H\left[\beta\gamma, \alpha\right]^{(u)} \right] \frac{\partial u_\beta}{\partial v_\mu} \frac{\partial u_\gamma}{\partial v_\nu}$$

$$= \sum_\beta \sum_\gamma H\left\{\beta\gamma, \epsilon\right\}^{(u)} \frac{\partial u_\beta}{\partial v_\mu} \frac{\partial u_\gamma}{\partial v_\nu}.$$

And

$$\sum_\beta \sum_\gamma H_{\beta\gamma}^{(u)} \frac{\partial^2 u_\beta}{\partial v_\mu \partial v_\nu} \sum_\lambda \sum_\rho H_{(v)}^{\lambda\rho} \frac{\partial u_\varepsilon}{\partial v_\rho} \frac{\partial u_\gamma}{\partial v_\lambda}$$

$$= \sum_{\beta} \left[\sum_{\gamma} H_{\beta\gamma}^{(u)} H_{(u)}^{\varepsilon\gamma} \right] \frac{\partial^2 u_\beta}{\partial v_\mu \partial v_\nu} = \frac{\partial^2 u_\varepsilon}{\partial v_\mu \partial v_\nu}.$$

Thus, transposing terms, the formula becomes

$$\frac{\partial^2 u_\epsilon}{\partial v_\mu \partial v_\nu} = \sum_{\rho} \frac{\partial u_\epsilon}{\partial v_\rho} H\{\mu\nu, \rho\}^{(v)} - \sum_{\beta} \sum_{\gamma} H\{\beta\gamma, \epsilon\}^{(u)} \frac{\partial u_\beta}{\partial v_\mu} \frac{\partial u_\gamma}{\partial v_\nu} \quad \text{.........(43)}.$$

This is the standard formula for $\dfrac{\partial^2 u_\epsilon}{\partial v_\mu \partial v_\nu}$.

Finally operate with

$$\sum_{\epsilon} \frac{\partial v_\lambda}{\partial u_\epsilon} *,$$

and transpose terms. We obtain (putting α for ϵ)

$$H\{\mu\nu, \lambda\}^{(v)} = \sum_{\alpha} \sum_{\beta} \sum_{\gamma} H\{\beta\gamma, \alpha\}^{(u)} \frac{\partial u_\beta}{\partial v_\mu} \frac{\partial u_\gamma}{\partial v_\nu} \frac{\partial u_\lambda}{\partial u_\alpha} + \sum_{\alpha} \frac{\partial^2 u_\alpha}{\partial v_\mu \partial v_\nu} \frac{\partial v_\lambda}{\partial u_\alpha} \quad \text{...........(44)}.$$

This is the standard formula relating the three-index symbols of the second type, as expressed in different coordinate-systems.

Now let $\|K_{\mu\nu}\|$ be another symmetric covariant tensor of the second order. Then we at once prove that

$$\left\| H\{\mu\nu, \lambda\}^{(v)} - K\{\mu\nu, \lambda\}^{(v)} \right\| \quad \text{.......................(45)}$$

is a mixed tensor of the third order, for which λ is the sole contravariant affix.

For, it at once follows from the formula above that

$$H\{\mu\nu, \lambda\}^{(v)} - K\{\mu\nu, \lambda\}^{(v)}$$
$$= \sum_{\alpha} \sum_{\beta} \sum_{\gamma} \left[H\{\beta\gamma, \alpha\}^{(u)} - K\{\beta\gamma, \alpha\}^{(u)} \right] \frac{\partial u_\beta}{\partial v_\mu} \frac{\partial u_\gamma}{\partial v_\nu} \frac{\partial v_\lambda}{\partial u_\alpha} \quad \text{...........(45·1)}.$$

This proves the theorem.

33. **Covariant Tensors of the First Order.** Let $\|T_\mu\|$ be a covariant tensor of the first order, and let $\|H_{\mu\nu}\|$ be any symmetric tensor of the second order. Then

$$T_\mu^{(v)} \sum_\alpha T_\alpha^{(u)} \frac{\partial u_\alpha}{\partial v_\mu},$$

and

$$\frac{\partial}{\partial v_\nu} = \sum_\beta \frac{\partial u_\beta}{\partial v_\nu} \frac{\partial}{\partial u_\beta}.$$

Hence

$$\frac{\partial T_\mu^{(v)}}{\partial v_\nu} = \sum_\alpha T_\alpha^{(u)} \frac{\partial^2 u_\alpha}{\partial v_\mu \partial v_\nu} + \sum_\alpha \sum_\beta \frac{\partial T_\alpha^{(u)}}{\partial u_\beta} \frac{\partial u_\alpha}{\partial v_\mu} \frac{\partial u_\beta}{\partial v_\nu}.$$

Now use formula (43), and remember that

$$\sum_\alpha \sum_\rho T_\alpha^{(u)} \frac{\partial u_\alpha}{\partial v_\rho} H\{\mu\nu, \rho\}^{(v)} = \sum_\rho T_\rho^{(v)} H\{\mu\nu, \rho\}^{(v)}.$$

We deduce

$$\frac{\partial T_\mu^{(v)}}{\partial v_\nu} - \sum_\rho T_\rho^{(v)} H\{\mu\nu, \rho\}^{(v)}$$

$$= \sum_\beta \sum_\gamma \left[\frac{\partial T_\beta^{(u)}}{\partial u_\gamma} - \sum_\alpha T_\alpha^{(u)} H\{\beta\gamma, \alpha\}^{(u)} \right] \frac{\partial u_\beta}{\partial v_\mu} \frac{\partial u_\gamma}{\partial v_\nu}.$$

Thus

$$\left\| \frac{\partial T_\mu^{(v)}}{\partial v_\nu} - \sum_\rho T_\rho^{(v)} H\{\mu\nu, \rho\}^{(v)} \right\| \quad \ldots\ldots\ldots\ldots\ldots(46).$$

is a covariant tensor of the second order.

Interchanging μ and v, and subtracting, we find that

$$\left\| \frac{\partial T_\mu^{(v)}}{\partial v_\nu} - \frac{\partial T_\nu^{(v)}}{\partial v_\mu} \right\| \quad \ldots\ldots\ldots\ldots\ldots\ldots\ldots\ldots\ldots(47)$$

is a covariant tensor of the second order.

When $\|T_\mu\|$ is a given covariant tensor of the first order, we shall use $T_{\mu\nu}$ to mean

$$T_{\mu\nu}^{(v)} = \frac{\partial T_{\mu}^{(v)}}{\partial v_{\nu}} - \frac{\partial T_{\nu}^{(v)}}{\partial v_{\mu}} \quad \text{.........................(47·1).}$$

Thus, with this meaning, $\|T_{\mu\nu}^{(\mu)}\|$ is a skew covariant tensor of the second order. Also identically

$$\frac{\partial T_{\mu\nu}^{(v)}}{\partial v_{\lambda}} + \frac{\partial T_{\nu\lambda}^{(v)}}{\partial v_{\mu}} + \frac{\partial T_{\lambda\mu}^{(v)}}{\partial v_{\nu}} = 0 \quad \text{.......................(47·2)}$$

34. **Contravariant Tensors of the First Order.** Let $\|S^{\mu}\|$ be any contravariant tensor of the first order. Then

$$S_{(v)}^{\mu} = \sum_{\alpha} S_{(u)}^{\alpha} \frac{\partial v_{\mu}}{\partial u_{\alpha}},$$

and

$$\frac{\partial}{\partial v_{\nu}} = \sum_{\beta} \frac{\partial u_{\beta}}{\partial v_{\nu}} \frac{\partial}{\partial u_{\beta}}.$$

Thus

$$\frac{\partial S_{(v)}^{\mu}}{\partial v_{\nu}} = \sum_{\alpha}\sum_{\beta} \frac{\partial S_{(u)}^{\alpha}}{\partial u_{\beta}} \frac{\partial v_{\mu}}{\partial u_{\alpha}} \frac{\partial u_{\beta}}{\partial v_{\nu}} + \sum_{\alpha}\sum_{\beta} S_{(u)}^{\alpha} \frac{\partial u_{\beta}}{\partial v_{\nu}} \frac{\partial^2 v_{\mu}}{\partial u_{\alpha} \partial u_{\beta}}.$$

Hence [cf. equation (43)]

$$\frac{\partial S_{(v)}^{\mu}}{\partial v_{\nu}} = \sum_{\alpha}\sum_{\beta} \frac{\partial S_{(u)}^{\alpha}}{\partial u_{\beta}} \frac{\partial v_{\mu}}{\partial u_{\alpha}} \frac{\partial u_{\beta}}{\partial v_{\nu}} + \sum_{\alpha}\sum_{\beta}\sum_{\epsilon} S_{(u)}^{\alpha} \frac{\partial v_{\mu}}{\partial u_{\epsilon}} \frac{\partial u_{\beta}}{\partial v_{\nu}} H\{\alpha\beta, \epsilon\}^{(u}$$

$$- \sum_{\rho}\sum_{\sigma} \left[\sum_{\alpha} S_{(u)}^{\alpha} \frac{\partial v_{\rho}}{\partial u_{\alpha}} \right] H\{\rho\sigma, \mu\}^{(v)} \left[\sum_{\beta} \frac{\partial u_{\beta}}{\partial v_{\nu}} \frac{\partial v_{\sigma}}{\partial u_{\beta}} \right].$$

Now in the second term on the right-hand side interchange α and ϵ, and in the third term note that

$$\sum_{\alpha} S_{(u)}^{\alpha} \frac{\partial v_{\rho}}{\partial u_{\alpha}} = S_{(v)}^{\rho}$$

and

$$\sum_{\beta} \frac{\partial u_{\beta}}{\partial v_{\nu}} \frac{\partial v_{\sigma}}{\partial u_{\beta}} = I_{\nu}^{\sigma}.$$

Hence, rearranging terms,

$$\frac{\partial S^{\mu}_{(v)}}{\partial v_{\nu}} + \sum_{\rho} S^{\rho}_{(v)} H\{\rho\nu, \mu\}^{(v)}$$

$$= \sum_{\alpha}\sum_{\beta}\left[\frac{\partial S^{\alpha}_{(u)}}{\partial u_{\beta}} + \sum_{\epsilon} S^{\epsilon}_{(u)} H\{\varepsilon\beta, \alpha\}^{(u)}\right]\frac{\partial v_{\mu}}{\partial u_{\alpha}}\frac{\partial u_{\beta}}{\partial v_{\nu}}.$$

Thus

$$\left\| \frac{\partial S^{\alpha}_{(u)}}{\partial u_{\beta}} + \sum_{\epsilon} S^{\epsilon}_{(u)} H\{\epsilon\beta, \alpha\}^{(u)} \right\| \quad \text{.................(48).}$$

is a mixed tensor of the second order, β being the covariant affix. It will be noticed that the differentiation adds the covariant affix to the original tensor.

Since we have a mixed tensor we can apply the process of restriction, identifying α and β. Hence [cf. equation (41) and changing ϵ to α]

$$\sum_{\alpha}\left[\frac{\partial S^{\alpha}_{(u)}}{\partial u_{\beta}} + S^{\alpha}_{(u)}\frac{\partial}{\partial u_{\alpha}}\log\left\{-H^{(u)}\right\}^{\frac{1}{2}}\right] \quad \text{.................(48.1)}$$

is invariant.

35. **An Example.** Let A be any scalar function of the position of an event-particle. Then $\left\|\frac{\partial A}{\partial u_{\alpha}}\right\|$ is a covariant tensor. Hence, using the Galilean tensor for the Cartesian group of coordinate-systems,

$$\left\| \sum_{\beta} G^{\alpha\beta}_{(u)} \frac{\partial A}{\partial u_{\beta}} \right\| \quad \text{.....................................(49)}$$

is a contravariant tensor. Hence [cf. formula (48·1) of section 34]

$$\sum_{\alpha}\sum_{\beta}\left[\frac{\partial}{\partial u_{\alpha}}\left\{G^{\alpha\beta}_{(u)}\frac{\partial A}{\partial u_{\beta}}\right\} + G^{\alpha\beta}_{(u)}\frac{\partial A}{\partial u_{\beta}}\frac{\partial}{\partial u_{\alpha}}\log\left\{-G^{(u)}\right\}^{\frac{1}{2}}\right] \quad \text{.....(49.1)}$$

is invariant.

In any Cartesian coordinate-system 'x' this invariant reduces to

$$-\left(\frac{\partial^2 A}{\partial x_1^{\;2}}+\frac{\partial^2 A}{\partial x_2^{\;2}}+\frac{\partial^2 A}{\partial x_2^{\;3}}\right)+\frac{1}{c^2}\frac{\partial^2 A}{\partial x_4^{\;2}} \text{.................} (49.2).$$

Thus we have transformed this fundamental expression to any coordinates.

Again substituting the covariant tensor $\left\|\frac{\partial A}{\partial v_\mu}\right\|$ for $\|T_\mu^{(v)}\|$ in formula (46) of section 33, we deduce that

$$\left\|\frac{\partial^2 A}{\partial v_\mu \partial v_\nu}-\sum_\rho \frac{\partial A}{\partial v_\rho} H\{\mu\nu,\rho\}^{(v)}\right\| \text{......................} (50)$$

is a covariant tensor of the second order.

36. **Tensors of the Second Order.** Any tensor of the second order can be expressed as a sum of products of pairs of tensors of the first order [cf. section 11].

Thus if $\|S_{\mu\nu}\|$ be a covariant tensor of the second order, we can write

$$\|S_{\mu\nu}\| = \|\Sigma\, A_\mu B_\nu\|,$$

where $\|A_\mu\|$ and $\|B_\nu\|$, etc., are covariant tensors of the first order.

Thus

$$\begin{aligned}\frac{\partial S_{\mu\nu}^{(u)}}{\partial u_\lambda} &= \Sigma\left[\frac{\partial A_\mu^{(u)}}{\partial u_\lambda} B_\nu^{(u)} + B_\mu^{(u)}\frac{\partial B_\nu^{(u)}}{\partial u_\lambda}\right]\\ &= \Sigma\left[\left(\frac{\partial A_\mu^{(u)}}{\partial u_\lambda}-\sum_\rho A_\rho^{(u)} H\{\lambda\mu,\rho\}^{(u)}\right)B_\nu^{(u)}\right.\\ &\qquad \left.+A_\mu^{(u)}\left(\frac{\partial B_\nu^{(u)}}{\partial u_\lambda}-\sum_\rho B_\rho^{(u)} H\{\lambda\mu,\rho\}^{(u)}\right)\right]\\ &\qquad +\sum_\rho\left(S_{\rho\nu}^{(u)} H\{\lambda\mu,\rho\}^{(u)} + S_{\mu\rho}^{(u)} H\{\lambda\nu,\rho\}^{(u)}\right).\end{aligned}$$

Hence [cf. formula (46)], since the sums of products of tensors are tensors,

$$\left\| \frac{\partial S_{\mu\nu}^{(u)}}{\partial u_\lambda} - \sum_\rho \left(S_{\rho\nu}^{(u)} H\{\lambda\mu, \rho\}^{(u)} + S_{\rho\nu}^{(u)} H\{\lambda\nu, \rho\}^{(u)} \right) \right\| \text{.........}(51)$$

is a covariant tensor of the third order, since it is equal to such a tensor.

Let $\|T^{\mu\nu}\|$ be a contravariant tensor of the second order. We can write

$$\left\| T^{\mu\nu} \right\| = \left\| \Sigma A^\mu B^\nu \right\|,$$

where $\|A_\mu\|$ and $\|B^\nu\|$ etc., are contravariant tensors of the first order. Thus

$$\frac{\partial T_{(u)}^{\mu\nu}}{\partial u_\lambda} = \Sigma \left[\frac{\partial A_{(u)}^\mu}{\partial u_\lambda} B_{(u)}^\nu + A_{(u)}^\mu \frac{\partial B_{(u)}^\nu}{\partial u_\lambda} \right]$$

$$= \Sigma \left[\left(\frac{\partial A_{(u)}^\mu}{\partial u_\lambda} + \sum_\rho A_{(u)}^\rho H\{\rho\lambda, \mu\}^{(u)} \right) B_{(u)}^\nu \right.$$

$$\left. + A_{(u)}^\mu \left(\frac{\partial B_{(u)}^\nu}{\partial u_\lambda} + \sum_\rho B_{(u)}^\rho H\{\rho\lambda, \nu\}^{(u)} \right) \right]$$

$$- \sum_\rho \left(T_{(u)}^{\rho\nu} H\{\rho\lambda, \mu\}^{(u)} + T_{(u)}^{\mu\rho} H\{\rho\lambda, \nu\}^{(u)} \right).$$

Hence [cf. formula (48)], since the sums of products of tensors are tensors,

$$\left\| \frac{\partial T_{(u)}^{\mu\nu}}{\partial u_\lambda} + \sum_\rho \left(T_{(u)}^{\rho\nu} H\{\rho\lambda, \mu\}^{(u)} + T_{(u)}^{\mu\rho} H\{\rho\lambda, \nu\}^{(u)} \right) \right\| \text{.......}(52)$$

is a mixed tensor of the third order, λ being the sole covariant affix.

Identifying λ and v and summing, we obtain by this process of restriction the tensor

$$\left\| \sum_\lambda \frac{\partial T_{(u)}^{\mu\lambda}}{\partial u_\lambda} + \sum_\rho \left(\sum_\lambda T_{(u)}^{\rho\lambda} H\{\rho\lambda, \mu\}^{(u)} + T_{(u)}^{\mu\rho} \frac{\partial}{\partial u_\rho} \log \left\{ -H^{(u)} \right\}^{\frac{1}{2}} \right) \right\|$$

$$\text{..................}(52.1),$$

which is a contravariant tensor of the first order.

Mixed tensors of the second order can be dealt with by exactly the same method as that applied to covariant and contravariant tensors in this article, and by the use of the same formulae (46) and (48). If $\|L_\mu^\nu(u)\|$ be a mixed tensor of the second order we deduce the tensor

$$\left\| \frac{\partial L_\mu^\nu(u)}{\partial u_\lambda} - \sum_\rho L_\mu^\nu(u) H\{\mu\lambda, \rho\}^{(u)} + \sum_\rho L_\mu^\rho(u) H\{\rho\lambda, \nu\}^{(u)} \right\|$$

which is a mixed tensor of the third order in which v is the sole contravariant affix.

Identifying λ and v and summing, we find by this process of restriction the tensor

$$\left\| \sum_\lambda \frac{\partial L_\mu^\nu(u)}{\partial u_\lambda} - \sum_\lambda \sum_\rho L_\mu^\lambda(u) H\{\mu\lambda, \rho\}^{(u)} \right.$$
$$\left. + \sum_\rho L_\mu^\rho(u) \frac{\partial}{\partial u_\rho} \log\left\{-H^{(u)}\right\}^{\frac{1}{2}} \right\|$$

which is a contravariant tensor of the first order.

37. **Tensors of the Third Order.** These are dealt with by the same method as those of the second order, by the use of the formulae obtained in sections 33, 34 and 36. The only such tensor which we need explicitly consider is a mixed tensor of the third order with only one contravariant affix. Let $\|K_{\mu\nu}^\lambda\|$ be such a tensor. We can write this tensor in the form

$$\left\| K_{\mu\nu}^\lambda \right\| = \left\| \Sigma\, A_{\mu\nu} B^\lambda \right\|,$$

where $\|A_{\mu\nu}\|$ is a covariant tensor of the second order and $\|B^\lambda\|$ is a contravariant tensor of the first order, and so on for the other pairs of tensors.

Hence by the use of formulae (51) and (48) we deduce that

$$\left\| \left[\frac{\partial K^{\lambda}_{\mu\nu}}{\partial u_{\pi}} + \sum_{\rho} K^{\rho}_{\mu\nu} H\{\rho\pi, \lambda\}^{(u)} \right. \right.$$
$$\left. \left. - \sum_{\rho} \left(K^{\lambda}_{\rho\nu} H\{\mu\pi, \rho\}^{(u)} + K^{\lambda}_{\rho\nu} H\{\nu\pi, \rho\}^{(u)} \right) \right] \right\| \cdots\cdots\cdots\cdots\cdots\cdots (54)$$

is a mixed tensor of the fourth order in which λ is the sole contravariant affix.

Identifying λ and (and summing, we obtain by this process of restriction the tensor

$$\left\| \left[\sum_{\lambda} \frac{\partial K^{\lambda}_{\mu\nu}}{\partial u_{\lambda}} + \sum_{\rho} K^{\rho}_{\mu\nu} \frac{\partial}{\partial u_{\rho}} \log\left\{-H^{(u)}\right\}^{\frac{1}{2}} \right. \right.$$
$$\left. \left. - \sum_{\lambda} \sum_{\rho} \left(K^{\lambda}_{\rho\nu} H\{\mu\lambda, \rho\}^{(u)} + K^{\lambda}_{\mu\rho} H\{\nu\lambda, \rho\}^{(u)} \right) \right] \right\|, \cdots\cdots\cdots\cdots\cdots (55)$$

which is a covariant tensor of the second order.

CHAPTER XXIV

SOME IMPORTANT TENSORS

38. **The Riemann-Christoffel Tensor.** Consider the Tensor differentiation of the covariant second-order tensor

$$\left\| \frac{\partial T_\mu^{(u)}}{\partial u_\nu} - \sum_\rho T_\rho^{(u)} H\{\mu\nu, \rho\}^{(u)} \right\|$$

which is obtained as formula (46) in section 33. We use formula (51) of section 36, substituting the given tensor for $\|S_{\mu\nu}\|$. We deduce, after arranging the terms, the covariant tensor

$$\left\| \frac{\partial^2 T_\mu^{(u)}}{\partial u_\lambda \partial u_\nu} - \sum_\rho \left[\frac{\partial T_\rho^{(u)}}{\partial u_\lambda} H\{\mu\nu, \rho\}^{(u)} + \frac{\partial T_\rho^{(u)}}{\partial u_\nu} H\{\mu\nu, \rho\}^{(u)} \right.\right.$$
$$\left. + \frac{\partial T_\mu}{\partial u_\rho} H\{\nu\lambda, \rho\}^{(u)} \right] - \sum_\rho T_\rho^{(u)} \left[\frac{\partial}{\partial u_\lambda} H\{\mu\nu, \rho\}^{(u)} \right.$$
$$- \sum_\sigma H\{\sigma\nu, \rho\}^{(u)} H\{\mu\lambda, \sigma\}^{(u)}$$
$$\left.\left. - \sum_\sigma H\{\sigma\mu, \rho\}^{(u)} H\{\nu\lambda, \sigma\}^{(u)} \right] \right\| \quad \text{.........(56).}$$

Now interchange λ and v in this tensor and subtract the latter from the former tensor. We obtain the tensor

$$\left\| \sum_\rho T_\rho^{(u)} \left[\frac{\partial}{\partial u_\nu} H\{\mu\lambda, \rho\}^{(u)} - \frac{\partial}{\partial u_\lambda} H\{\mu\nu, \rho\}^{(u)} \right. \right.$$
$$+ \sum_\sigma H\{\sigma\nu, \rho\}^{(u)} H\{\mu\lambda, \sigma\}^{(u)}$$
$$\left. \left. - \sum_\sigma H\{\sigma\lambda, \rho\}^{(u)} H\{\mu\nu, \sigma\}^{(u)} \right] \right\|.$$

Hence by section 17,

$$\left\| \frac{\partial}{\partial u_\nu} H\{\mu\lambda, \pi\}^{(u)} - \frac{\partial}{\partial u_\lambda} H\{\mu\nu, \pi\}^{(u)} + \sum_\sigma H\{\sigma\nu, \pi\}^{(u)} H\{\mu\lambda, \sigma\}^{(u} \right.$$
$$\left. - \sum_\sigma H\{\sigma\lambda, \pi\}^{(u)} H\{\mu\nu, \sigma\}^{(u)} \right\| \quad \ldots\ldots\ldots(57).$$

is a mixed tensor of the fourth order, in which π is the only contravariant affix. This is the Riemann-Christoffel Tensor.

Now identify ν and π and sum. Then, as the result of this process of restriction, we obtain the covariant second-order tensor

$$\left\| \sum_\rho \frac{\partial}{\partial u_\rho} H\{\mu\lambda, \rho\}^{(u)} - \frac{\partial^2}{\partial u_\lambda \partial u_\mu} \log\left\{-H^{(u)}\right\}^{\frac{1}{2}} \right.$$
$$+ \sum_\sigma H\{\mu\lambda, \sigma\}^{(u)} \frac{\partial}{\partial u_\sigma} \log\left\{-H^{(u)}\right\}^{\frac{1}{2}}$$
$$\left. - \sum_\rho \sum_\sigma H\{\sigma\lambda, \rho\}^{(u)} H\{\mu\rho, \sigma\}^{(u)} \right\| \quad \ldots\ldots(58).$$

This is the restricted Hiemann-Christoffel Tensor. It is a symmetric covariant second-order tensor.

39. **The Linear Gravitational Tensor.** In formula (45) of section 32, we have proved that, if $\|H_{\mu\nu}\|$ and $\|J_{\mu\nu}\|$ are any symmetric covariant second-order tensors,

$$\left\| J\{\mu\nu, \lambda\}^{(u)} - H\{\mu\nu, \lambda\}^{(u)} \right\| \quad \ldots\ldots\ldots\ldots\ldots\ldots(59)$$

is a third-order mixed tensor in which λ is the sole contravariant index.

Hence operating with

$$\sum_{\lambda} J_{\pi\lambda}*$$

we find that

$$\left\| J\left[\mu\nu,\pi\right]^{(u)} - \sum_{\lambda} J_{\pi\lambda}^{(u)} H\left\{\mu\nu,\lambda\right\}^{(u)} \right\| \quad \text{................} (59{\cdot}1)$$

is a covariant third-order tensor.

Again operating on this latter tensor with

$$\sum_{\pi} H_{(u)}^{\rho\pi}*,$$

and interchanging the indices, viz. putting λ for *p*, *p* for *π*, and σ for λ, we find that

$$\left\| \sum_{\rho} H_{(u)}^{\lambda\rho} J\left[\mu\nu,\rho\right]^{(u)} - \sum_{\rho}\sum_{\sigma} H_{(u)}^{\lambda\rho} J_{\rho\sigma}^{(u)} H\left\{\mu\nu,\sigma\right\}^{(u)} \right\| \quad \text{.......} (59{\cdot}2)$$

is a third-order mixed tensor in which λ is the sole contravariant index.

Now we replace $\|H_{\mu\nu}\|$ by the Galilean tensor $\|G_{\mu\nu}\|$, and obtain the tensor

$$\left\| \sum_{\rho} G_{(u)}^{\lambda\rho} J\left[\mu\nu,\rho\right]^{(u)} - \sum_{\rho}\sum_{\sigma} G_{(u)}^{\lambda\rho} J_{\rho\sigma}^{(u)} G\left\{\mu\nu,\sigma\right\}^{(u)} \right\| \quad \text{......} (59{\cdot}3).$$

This tensor is linear in the components of $\|J_{\mu\nu}\|$. In any coordinate-system '*x*' of the Cartesian group, this tensor reduces to

$$\left\| -\frac{1}{\omega_{\lambda}^{2}} J\left[\mu\nu,\lambda\right]^{(x)} \right\| \quad \text{............................} (59{\cdot}4)$$

If in formula (55) of section 37, we replace $\|H_{\mu\nu}\|$ by $\|G_{\mu\nu}\|$ and replace $\|K_{\mu\nu}^{\lambda}\|$ by the above tensor, we obtain the tensor utilised in the law (ii) of gravitation mentioned in Chapter IV of Part I. In the coordinate-system '*x*' this law of gravitation becomes

$$\sum_{\lambda} \frac{1}{\omega_{\lambda}^{2}} \frac{\partial}{\partial x_{\lambda}} J\left[\mu\nu,\lambda\right]^{(x)} = 0, \quad \left[\mu,\nu = 1,2,3,4\right] \quad \text{...........} (60)$$

where there is no attracting matter.

It is evident that $\|H_{\mu\nu}\|$ is introduced in the above reasoning disconnectedly on two distinct occasions, namely in formula (59), and in the operation $\sum_{\pi} H_{(u)}^{\rho\pi}*$.

There is no logical necessity that $\|H_{\mu\nu}\|$ should be the same on each of these occasions, still less that it should be the Galilean tensor. Accordingly this is an opportunity of framing other laws of gravitation in which tensors characteristic of other fields of force are introduced instead of $\|G_{\mu\nu}\|$ on one or more of these three occasions. In this way, the influence (if any) of these fields on the gravitational field may be represented.

40. **Cyclic Reduction.** The Cyclic Reduction of the third-order array $\|A_{\mu\nu}^{\lambda}\|$ is the array

$$\left\| A_{\lambda\mu\nu} + A_{\mu\nu\lambda} + A_{\nu\lambda\mu} \right\| \text{..............................} (61).$$

This reduced array will be symbolised by

$$\left\| \text{Cycl. } A_{\lambda\mu\nu} \right\| \text{..............................} (61.1).$$

This definition and the symbolism will be applied to arrays of any order. Thus

$$\left\| \text{Cycl. } A_{\mu\nu} \right\| = \left\| A_{\mu\nu} + A_{\nu\mu} \right\| \text{......................} (61.2),$$

and

$$\left\| \text{Cycl. } A_{\lambda\mu\nu\pi} \right\| = \left\| A_{\lambda\mu\nu\pi} + A_{\mu\nu\pi\lambda} + A_{\nu\pi\lambda\mu} + A_{\pi\lambda\mu\nu} \right\| \text{.....} (61.3).$$

The cyclic reduction of a covariant, or contravariant, tensor of any order is a tensor of the same order and type as the given tensor. The permanence of the order is obvious; we have only to prove the tensor property. Consider a covariant tensor of the third order. Then

$$T_{\lambda\mu\nu}^{(v)} = \sum_{\alpha}\sum_{\beta}\sum_{\gamma} T_{\alpha\beta\gamma}^{(u)} \frac{\partial u_{\alpha}}{\partial v_{\lambda}} \frac{\partial u_{\beta}}{\partial v_{\mu}} \frac{\partial v_{\gamma}}{\partial v_{\nu}}.$$

Now permute (λ, μ, ν) and (α, β, γ) each cyclically. Then

$$T_{\mu\nu\lambda}^{(v)} = \sum_{\alpha}\sum_{\beta}\sum_{\gamma} T_{\beta\gamma\alpha}^{(u)} \frac{\partial u_{\beta}}{\partial v_{\mu}} \frac{\partial v_{\gamma}}{\partial v_{\nu}} \frac{\partial u_{\alpha}}{\partial v_{\lambda}},$$

and analogously for $T_{\nu\lambda\mu}^{(v)}$. Hence

$$\text{Cycl. } T_{\lambda\mu\nu}^{(v)} = \sum_{\alpha}\sum_{\beta}\sum_{\gamma}\left[\text{Cycl. } T_{\alpha\beta\gamma}^{(u)}\right]\frac{\partial u_\alpha}{\partial v_\lambda}\frac{\partial u_\beta}{\partial v_\mu}\frac{\partial u_\gamma}{\partial v_\nu}.$$

This proves the required proposition, and an analogous proof evidently holds for contravariance, and for tensors of any order.

Now replace the covariant tensor $\|T_{\lambda\mu\nu}\|$ by

$$\left\|\frac{\partial S_{\mu\nu}^{(v)}}{\partial v_\lambda} - \sum_\rho\left[S_{\rho\nu}^{(v)} H\{\mu\lambda, \rho\}^{(v)} + S_{\mu\rho}^{(v)} H\{\nu\lambda, \rho\}^{(v)}\right]\right\|.$$

The preceding theorem on reduction, applied to this case, tells us that

$$\left\|\left[\frac{\partial S_{\mu\nu}^{(v)}}{\partial v_\lambda} + \frac{\partial S_{\nu\lambda}^{(v)}}{\partial v_\mu} + \frac{\partial S_{\lambda\mu}^{(v)}}{\partial v_\nu} - \sum_\rho\left\{\left(S_{\rho\lambda}^{(v)} + S_{\lambda\rho}^{(v)}\right)H\{\mu\nu, \rho\}^{(v)}\right.\right.\right.$$
$$\left.\left.\left. + \left(S_{\rho\mu}^{(v)} + S_{\mu\rho}^{(v)}\right)H\{\nu\lambda, \rho\}^{(v)} + \left(S_{\rho\nu}^{(v)} + S_{\nu\rho}^{(v)}\right)H\{\lambda\mu, \rho\}^{(v)}\right\}\right]\right\| \quad \text{.......(62)}$$

is a covariant third-order tensor.

Hence if $\|S_{\mu\nu}\|$ be a skew tensor, then

$$\left\|\frac{\partial S_{\mu\nu}^{(v)}}{\partial v_\lambda} + \frac{\partial S_{\nu\lambda}^{(v)}}{\partial v_\mu} + \frac{\partial S_{\lambda\mu}^{(v)}}{\partial v_\nu}\right\| \quad \text{........................(62.1)}$$

is a covariant third-order tensor.

41. **Some Cartesian Group Tensors.** We first note that if $\|S_\mu\|$ and $\|T_\mu\|$ be Cartesian group tensors (covariant and contravariant) of the first order, then [cf. section 29]

$$\left\|\frac{1}{\omega_\mu^2} S_\mu\right\| \text{ and } \left\|\omega_\mu^2 T^\mu\right\| \quad \text{.........................(63)}$$

are Cartesian group tensors (contravariant and covariant) of the first order. Furthermore if $\|S_{\mu\nu}\|$ and $\|T^{\mu\nu}\|$ be Cartesian group tensors (covariant and contravariant) of the second order, then [cf. section 28]

$$\left\| \frac{1}{\omega_\mu^2 \omega_\nu^2} S_{\mu\nu} \right\| \text{and} \left\| \omega_\mu^2 \omega_\nu^2 T^{\mu\nu} \right\| \text{..............(63·1)}$$

are Cartesian group tensors (contravariant and covariant) of the second order. And so on for higher orders.

Also in the case of the two Cartesian tensors (covariant and contravariant) of the first order

$$\sum_\mu \frac{1}{\omega_\mu^2} \left\{ S_\mu \right\}^2 \text{ and } \sum_\mu \omega_\mu^2 \left\{ T^\mu \right\}^2 \text{..................(63.2)}$$

are Cartesian invariants. And in the case of the two Cartesian tensors of the second order

$$\sum_\mu \frac{1}{\omega_\mu^2} S_{\mu\nu} \text{ and } \sum_\mu \omega_\mu^2 T^{\mu\nu} \text{......................(63·3)}$$

are Cartesian invariants.

Let (x_1, x_2, x_3, x_4) and (p_1, p_2, p_3, p_4) be the coordinates of any two event-particles referred to the same coordinate-system 'x' of the Cartesian group. Then

$$\left\| x_\mu - p_\mu \right\| \text{and} \left\| dx_\mu \right\| \text{and} \left\| dp_\mu \right\| \text{......................(64)}$$

are contravariant Cartesian tensors of the first order.

Thus

$$\sum_\mu \omega_\mu^2 \left(x_\mu - p_\mu \right)^2 \text{ and } \sum_\mu \omega_\mu^2 \, dx_\mu^2 \text{ and } \sum_\mu \omega_\mu^2 \, dp_\mu^2 \text{ ....(64.1)}$$

are Cartesian invariants. Put

$$r = \sqrt{\sum_\mu{}' \left(x_\mu - p_\mu \right)^2} \text{..............................(65).}$$

Then $$\sum_\mu \omega_\mu^2 \left(x_\mu - p_\mu \right)^2 = r^2 - c^2 \left(x_4 - p_4 \right)^2 \text{ ....(65.1).}$$

Also put

$$\dot{x}_\mu = \frac{dx_\mu}{dx_4}, \quad \dot{p}_\mu = \frac{dp_\mu}{dp_4}, \quad [\mu = 1, 2, 3, 4] \quad \text{..........(65·2)},$$

$$v_M^2 = \dot{x}_1^{\,2} + \dot{x}_2^{\,2} + \dot{x}_3^{\,2}, \quad v_m^2 = \dot{p}_1^2 + \dot{p}_2^2 + \dot{p}_3^2 \quad \text{......(65·3)},$$

$$\Omega_M = \left\{1 - \frac{v_M^2}{c^2}\right\}^{-\frac{1}{2}}, \quad \Omega_m = \left\{1 - \frac{v_m^2}{c^2}\right\}^{-\frac{1}{2}} \quad \text{.........(65·4)},$$

$$\xi_m = \frac{1}{c}\sum_\mu{}' \left(x_\mu - p_\mu\right)\dot{p}_\mu \quad \text{.................(65·5)};$$

then from above

$$\Omega_M^{-1} dx_4 \text{ and } \Omega_M^{-1} dp_4 \quad \text{..............................(66)}$$

are Cartesian invariants.

Hence differentiating the Cartesian tensor $x^\mu - p^\mu$ with respect to x_4, we deduce that

$$\left\| \Omega_M \dot{x}_\mu \right\| \text{ and } \left\| \omega_\mu^2 \Omega_M \dot{x}_\mu \right\| \quad \text{.......................(67)}$$

are Cartesian tensors (contravariant and covariant). Also differentiating the Cartesian invariant $r^2 - c^3\ (x_4 - p_4)^2$ with respect to p_4, we deduce that

$$\Omega_m \left\{c\left(x_4 - p_4\right) - \xi_m\right\} \quad \text{.............................(68)}$$

is a cartesian invariant.

Again differentiating the Cartesian tensor with respect to x_4, we deduce that

$$\left\| \Omega_M^2 \left\{ \ddot{x}_\mu + \frac{\Omega_M^2}{c^2} \dot{x}_\mu \sum_\rho{}' \dot{x}_\rho \ddot{x}_\rho \right\} \right\|$$

$$\text{and } \left\| \omega_\mu^2 \Omega_M^2 \left\{ \ddot{x}_\mu + \frac{\Omega_M^2}{c^2} \dot{x}_\mu \sum_\rho{}' \dot{x}_\rho \ddot{x}_\rho \right\} \right\| \quad \text{....(69)}$$

are Cartesian tensors (contravariant and covariant).

Again differentiating the Cartesian invariant

$$\Omega_m\left\{c\left(x_4 - p_4\right) - \xi_m\right\}$$

with respect to x_4 we find that

$$\Omega_m \Omega_M \left\{ 1 - \frac{1}{c^2} \sum_\mu{}' \dot{x}_\mu \dot{p}_\mu \right\} \qquad \ldots\ldots\ldots\ldots\ldots\ldots\ldots\ldots (70)$$

is a Cartesian invariant.

Also differentiating the same Cartesian invariant with respect to p_4, we find that

$$\frac{1}{c} \Omega_m^4 \left\{c\left(x_4 - p_4\right) - \xi_m\right\} \sum_\mu{}' \dot{p}_\mu \ddot{p}_\mu - \Omega_m^2 \sum_\mu{}' \left(x_\mu - p_\mu\right) \ddot{p}_\mu \quad \ldots\ldots (71)$$

is a Cartesian invariant.

Also from either of the tensors of formula (69) we find that

$$\Omega_M^4 \left[\sum_\mu{}' \ddot{x}_\mu^2 + \frac{1}{c^2} \Omega_M^2 \left(1 + \Omega_M^2\right) \left(\sum_\mu{}' \dot{x}_\mu \ddot{x}_\mu\right)^2 \right] \ldots\ldots\ldots\ldots (72)$$

is a Cartesian invariant.

ENDNOTES

PREFACE

[1] *The Principles of Natural Knowledge, and The Concept of Nature,* both Cambridge Univ. Press.

[2] June 5, 1922.

[3] April 18, 1922.

CHAPTER I

[1] Dec. 30, 1920

[2] *The Principles of Natural Knowledge,* and *The Concept of Nature,* both Camb. Univ. Press.

CHAPTER II

[1] I borrow the term 'historical' from Prof. C. D. Broad.

CHAPTER IV

[1] In Part II the 'Limb Effect' and the doubling or trebling of the spectral lines are also deduced.

[2] Cf. Part II, Chapter v, equation (8), and Chapter vi, equation (13).

CHAPTER VI

[1] The definition of $J^{\mu\nu}_{(u)}$ and of $G^{\mu\nu}_{(u)}$ is given in Chapter x, equation (3), below, and that of $G[\mu v, p]^{(\mu)}$ in Chapter v, equation (8), above.

SUGGESTED READING

BONDI, HERMANN. *Relativity and Common Sense.* New York: Dover Publications, Inc., 1980.

EARMAN, JOHN. *World Enough and Space-time.* Cambridge, MA: MIT Press, 1989.

EINSTEIN, ALBERT. *Ideas and Opinions.* New York: Crown, 1954.

---. *Relativity: The Special and the General Theory.* New York: Barnes and Noble Books, 2004.

EINSTEIN, ALBERT, H. A. LORENTZ, H. MINKOWSKI, AND H. WEYL. *The Principle of Relativity.* New York: Dover Publications, Inc., 1952.

FRIEDMAN, MICHAEL. *Foundations of Space-Time Theories.* Princeton, NJ: Princeton University Press, 1983.

GEROCH, ROBERT. *General Relativity from A to B.* Chicago, IL: University of Chicago Press, 1978.

HUGGET, NICK, ED. *Space from Zeno to Einstein.* Cambridge, MA: MIT Press, 1999.

JANOSSY, L. *The Theory of Relativity Based on Physical Reality.* Budapest: Academiaia Kiado, 1971.

REICHENBACH, HANS. *The Philosophy of Space and Time.* Trans. Maria Reichenbach. New York: Dover Publications, Inc., 1958.

TORRETTI, ROBERTO. *Relativity and Geometry.* New York: Dover Publications, Inc., 1983.

WHITEHEAD, A. N. *Process and Reality.* Cambridge: Cambridge University Press, 1928.

---.*Science and the Modern World.* Cambridge: Cambridge University Press, 1925.

---. *The Concept of Nature.* Cambridge: Cambridge University Press, 1920.
WILL, C. M. *Theory and Experiment in Gravitational Physics.* Cambridge: Cambridge University Press, 1993.